W9-BKN-542

A LITTLE BOOK OF COINCIDENCE

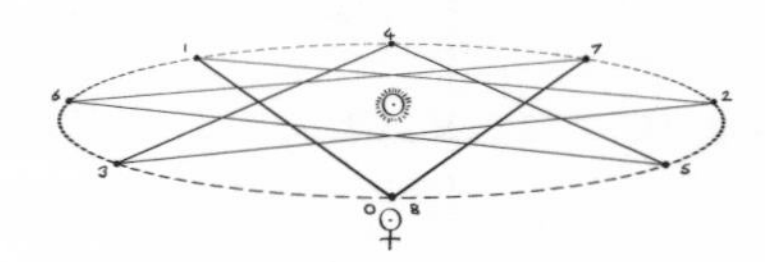

Originally published in Wales by Wooden Books Ltd. in 2001; first published in the United States of America in 2002 by Walker Publishing Company, Inc.

Published simultaneously in Canada by Fitzhenry and Whiteside, Markham, Ontario L3R 4T8

For information about permission to reproduce selections from this book, write to Permissions, Walker & Company, 435 Hudson Street, New York, New York 10014

Printed on recycled paper.

Library of Congress Cataloging-in-Publication Data

Martineau, John.

A little book of coincidence / written and illustrated by John Martineau.

p.cm.

Originally published: Presteigne, Powys, Wales: Wooden Books, 1995.

ISBN 0-8027-1388-2 (alk. paper)

1. Cosmology—Miscellanea. 2. Astronomy—Miscellanea. 3. Geometry—Miscellanea. 4. Coincidence. I. Title.

BD701 M37 2002

113—dc21

2001055919

Visit Walker & Company's Web site at www.walkerbooks.com

Printed in the United States of America

2 4 6 8 10 9 7 5 3 1

A LITTLE BOOK OF COINCIDENCE

written and illustrated by

John Martineau

Walker & Company
New York

To people who have tragically grown up
in a world devoid of a magical cosmology

Thanks to the numerous friends, colleagues, and others who have contributed to this project over the years. Please keep crunching the numbers and sending ideas and miracles.

Note: Percentages in brackets throughout the text refer to accuracies.

Early visions of an infinite universe of solar systems hinted at repeated structures like galaxies and parallel universes. From Thomas Wright's The Cosmos, *1750.*

Illustration on page i: *Venus's positions over eight birthdays draw an octagram around the Sun.*

Contents

A useful set of glyphs for the planets. Drawn by calligrapher Mark Mills, each made from Sun, Moon, and Earth, they are used throughout this book.

Introduction

Biological life is now thought to have appeared on this planet not long after its formation. It seems that the bacterial seeds for the process may have flown in on the tail of a comet or meteor. Speculation is again rife about life under the surface of Mars, on Jupiter's icy moon Europa, and, indeed, anywhere the sacred substance of liquid water is known to exist.

The science of the cosmos has changed immeasurably since the Greek and medieval visions of circles of planetary spheres. But with great cosmic schemes out of fashion, and with dragons and unicorns dismissed, the Earth has become a modern mystery. No modern theory exists to explain the miracle of conscious life nor the cosmic coincidences that surround our planet. Why do the Sun and Moon appear to be the same size in the sky? There are ancient answers to such questions, however, and these invoke liberal arts like music and geometry.

This book is not just another pocket guide to our solar system, for it suggests there may be fundamental relationships between space, time, and life that have not yet been understood. These days we scan the skies listening for intelligent radio signals and looking for remote planets a little like our own. Meanwhile, our closest planetary neighbors are making the most exquisite patterns around us, in space and in time, and no scientist has yet explained why. Is it all just a coincidence or do the patterns perhaps explain the scientists . . .

Radnorshire, May 2001

GALACTIC DUST
the well-tuned universe

There's a lot going on in the universe. As many star-filled galaxies pepper the bubble of Earth's space-time horizon as there are grains of sand on a beach. Our planet and we ourselves are made from reorganized, smoky stardust, a fact long taught by ancient cultures. We now know that stardust itself is made simply from fizzballs, organized whirlpools of light, long ago squeezed together deep inside stars. We live in between the little and the large, in a time and a place in the universe where things have condensed, crystallized, built up, and settled down.

Science still doesn't know whether conscious life is rare or common in the universe. Just how special are we and our Earth? Funnily enough, scientists are currently puzzling over the strange fact that the whole universe seems special. There is exactly enough material in the universe to make it, and the ratios between the fundamental forces seem specifically tuned to produce an amazingly complex, beautiful, and enduring universe. Slightly fiddle with any bit of it and you get a universe of black holes, insubstantial fizzballs, or other lifeless setups. Is this design or coincidence?

The story of the search for order, pattern, and meaning in the cosmos is very old indeed. The planets of our solar system have long been suspected of hiding secret relationships. In antiquity students of such things pondered the "Music of the Spheres," the heavenly bodies singing subtle and perfect harmonies to the adept. Today we have the simple precision of Kepler's, Newton's, and Einstein's laws. Who knows what will come next?

YOU ARE
HERE

The Solar System
spirals everywhere

It is now thought that our solar system condensed from a disk of debris some five billion years ago to form a Sun. Remaining heavier materials were attracted to each other and pulled inward to form small asteroids and rocky planets. Lighter gases were blown farther out by the solar wind to condense as the four gas giants, Jupiter, Saturn, Neptune, and Uranus. In the inner solar system, asteroids grew into planets, pulling the final pieces into place with more and more energy (today many remain hot inside from the collisions). Eventually things became as they are now.

The plane of the solar system is tilted at 30° to the plane of the galaxy so our solar system actually corkscrews its way around the arm of the Milky Way. The picture (*opposite, above; after Windelius and Tucker*) is a schematic of the motions of the four inner planets.

Another way to picture the solar system is by thinking of space-time as a rubber sheet with the Sun as a heavy ball and planets as marbles placed on it (*opposite, below; after Guy Murchie*). This is Einstein's model of the way matter curves space-time and helps visualize the force of gravity between masses. If we flick a frictionless pea onto our sheet, it could easily be captured by one of the marbles, or be spun around a few times and spat out, or settle into a fast spinning elliptical orbit halfway down any one of the wormholes. Like a planet, the farther the pea gets down the funnel, the faster it must circle to stop itself from going down the tube. Also, the faster it spins the heavier it gets and the slightly slower its clocks seem to run.

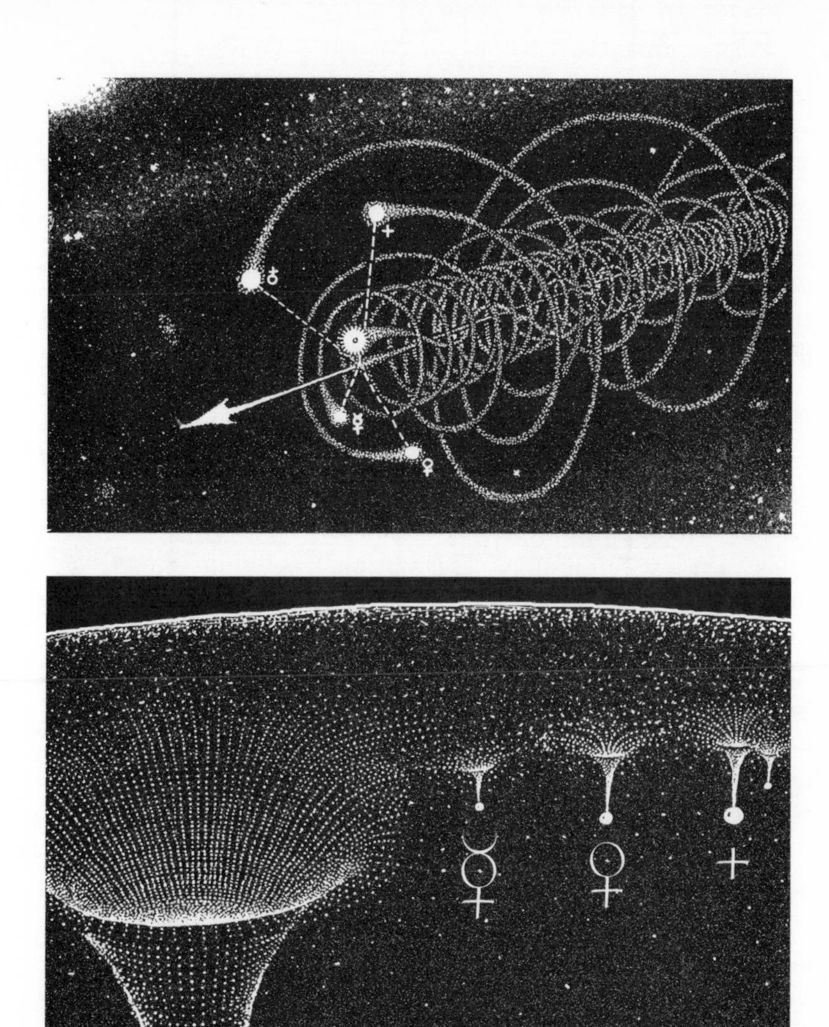

Retrograde Motion
running around kissing

Anyone watching the sky with the naked eye from Earth will notice that, apart from the steadily moving Sun and Moon, there are five *wandering* stars: the five planets of antiquity. These, and the newly discovered planets, appear to move around the Earth roughly following the Sun's annual circle, the *ecliptic* or *zodiac.* If only life was this simple! Watch planets for any length of time, and, far from moving in any simple way, they lurch around like drunken bees, waltzing and whirling. Occasionally, when planets pass, or kiss, each appears to the other to *retrogress,* or go backward, against the stars for a time. This was once common knowledge.

Below we see Mercury's pattern around a tracked Sun over a year as seen from Earth (*after Joachim Schultz*). Opposite is shown Cassini's early eighteenth-century sketch of the movements of Jupiter and Saturn. In ancient times hugely complex systems of circles and wheels were called into play to try to mimic these planetary motions (*opposite, below*), culminating in the Ptolemaic system of thirty-nine deferents and epicycles, used to model the motions of the seven heavenly bodies over two thousand years ago.

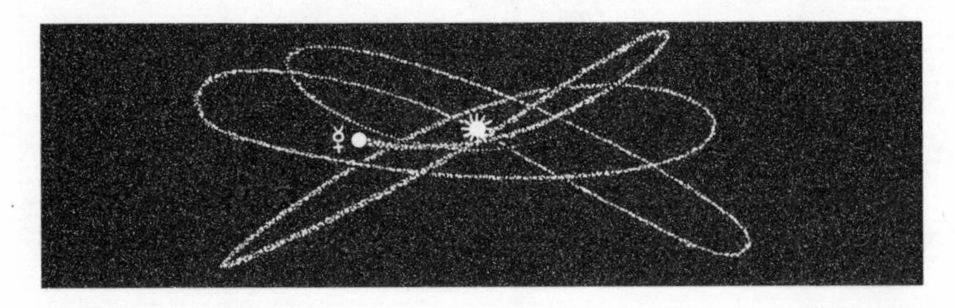

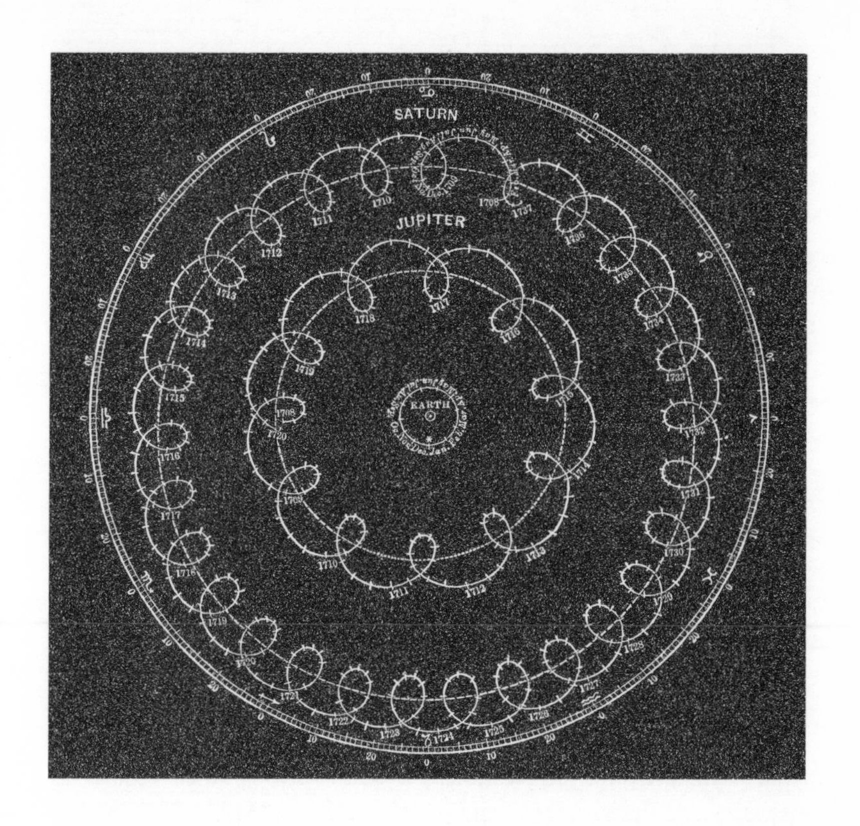

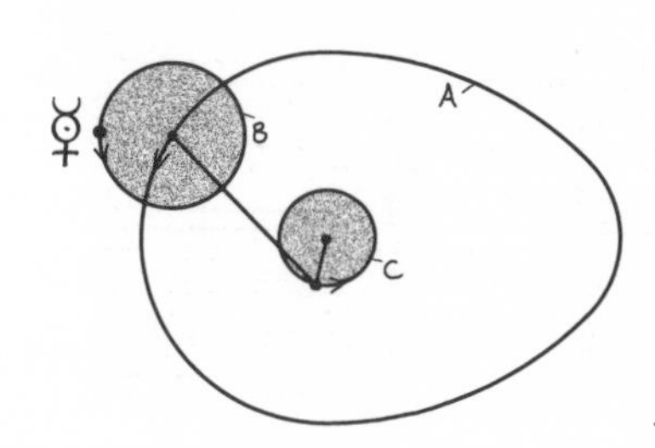

Until the 1600s all planetary motions were modeled using a "deferent" (A) basically circular, drawn from a central "eccentric" around which an "epicycle" moved (B), itself rotating with a planet attached. Tricks refined the system; here a kind of spinning crank (C) called a "movable eccentric" produces an egg-shaped deferent for Mercury's apparent dance in the heavens.

The Ancient Secret of Sevens
planets, metals, and days of the week

A short four hundred years ago the diagrams opposite formed the cornerstone of cosmological thought across the western world, as they had done for many thousands of years. Today, these emblems of the sevenfold system of antiquity appear as quaint reminders of an alchemical cosmology now buried beneath newly discovered planets and physical elements. Let us, however, take a quick look at the cosmology of our ancestors and see what it can teach us.

There are seven clearly visible moving heavenly bodies, and they may be arranged around a heptagon in the order of their apparent speed against the fixed stars. The Moon appears to move the fastest, followed by Mercury, Venus, the Sun, Mars, Jupiter, and Saturn (*top left*). Each heavenly body was then assigned to a day, still clear in many languages. The order of the days was given by the heptagram shown (*top right*). In English, older names for the planets (or gods) were used, thus Wotan's day, Thor's day, and Freya's day.

In antiquity seven metals were held to correspond with the seven planets, their compounds giving rise to color associations. Venus, for example, was associated with the greens and blues of copper carbonates. Students of alchemy would often ponder these magical relationships as they forged ever more subtle things. Astonishingly, the ancient system also gives the modern order by *atomic number* of these metals. Follow a more open heptagram to give: iron 26, copper 29, silver 49, tin 50, gold 79, mercury 80, and lead 82 (*lower left; after Critchlow and Hinze*). The electrical conductivity sequence also appears around the outside starting with lead.

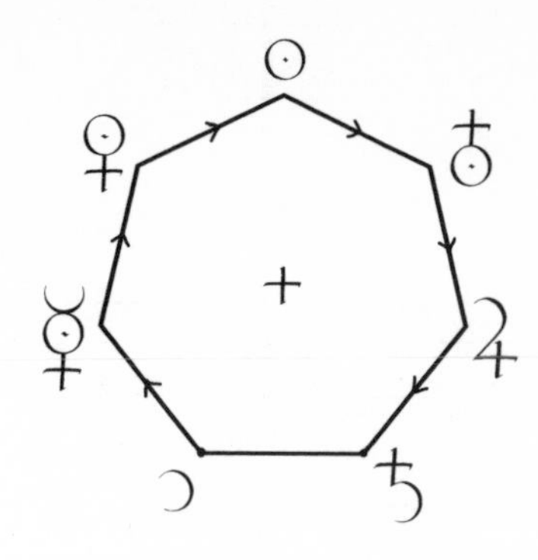

The Seven Heavenly Bodies
Start at the Moon and follow the arrows to give the "Chaldean Order" of the spheres.

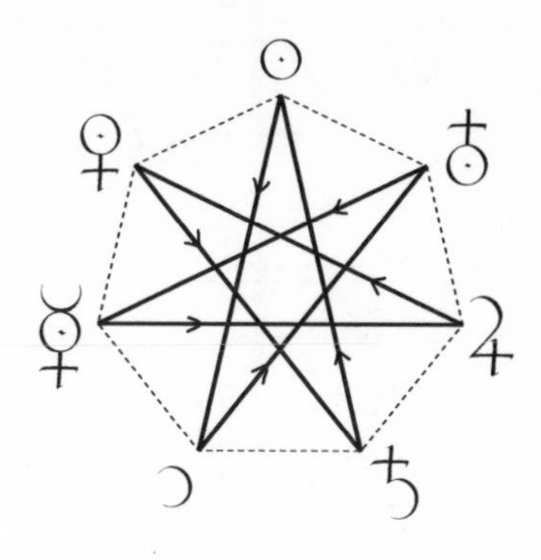

The Seven Days of the Week
French: Lundi, Mardi, Mercredi, Jeudi, Vendredi . . . follow the arrows again.

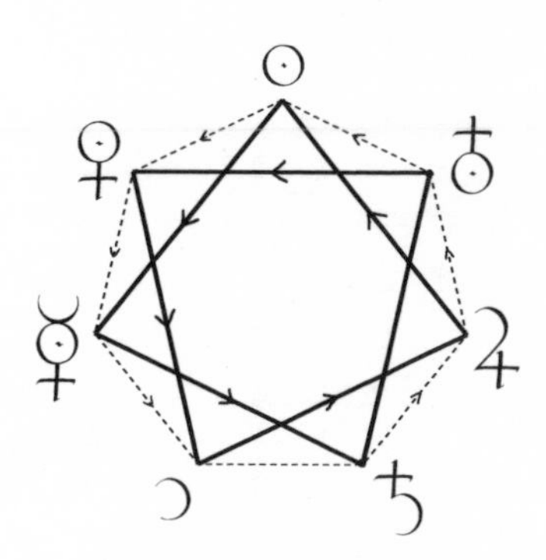

The Seven Metals of Antiquity
Start with iron and follow the arrows to give elements of increasing atomic number.

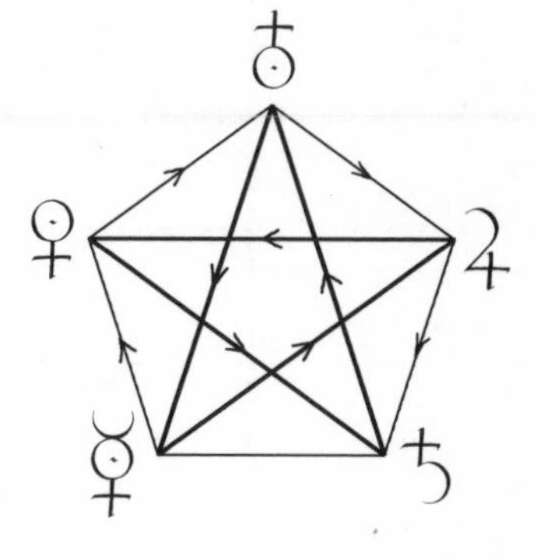

Removing the Sun and Moon
Start with Mercury. Moving around the pentagon now gives increasing distance from the Sun.

GEOCENTRIC OR HELIOCENTRIC

Earth or Sun at the center

The extraordinary Ptolemaic world of epicycles and deferents lasted a surprisingly long time. Despite its complexity it "saved appearances" and was even held to save souls. Ellipses were, in fact, studied by early Greek mathematicians such as Appollonius, and as early as 250 B.C. Aristarchus of Samos was proposing a system of planets orbiting the Sun. However, it was not to be, and for one and a half thousand years the Earth remained as we still experience it—a motionless body in the center of the universe, surrounded by whirling circles. The Ptolemaic system was handed down from the Greeks to the Arabs and then finally back to the West again.

Four ancient systems are shown opposite (*after Arthur Koestler*), and each sphere of each diagram is to be understood as having its own attachment of epicycles and eccentrics (*see page* 7). Copernicus, despite in 1543 placing the Sun in the center (*opposite, top left*), remained a convinced epicycle man, increasing the number of invisible wheels from the Ptolemaic thirty-nine up to an amazing forty-eight. In the late sixteenth century Tycho Brahe, against Kepler's evidence, desperately tried to keep the Earth stationary in the center of the universe (*opposite, bottom left*), while an early Greek model, Herakleides', like a later version by Eriugina, attempted a compromise.

The modern model of the solar system is shown at the bottom of page 11. It shows the planets (including the largest asteroid, Ceres) orbiting the Sun in space. Each planet has an orbital shell, some thicker than others. This basic model was first conceived by Johannes Kepler in 1596, and it is to his ideas that we now turn.

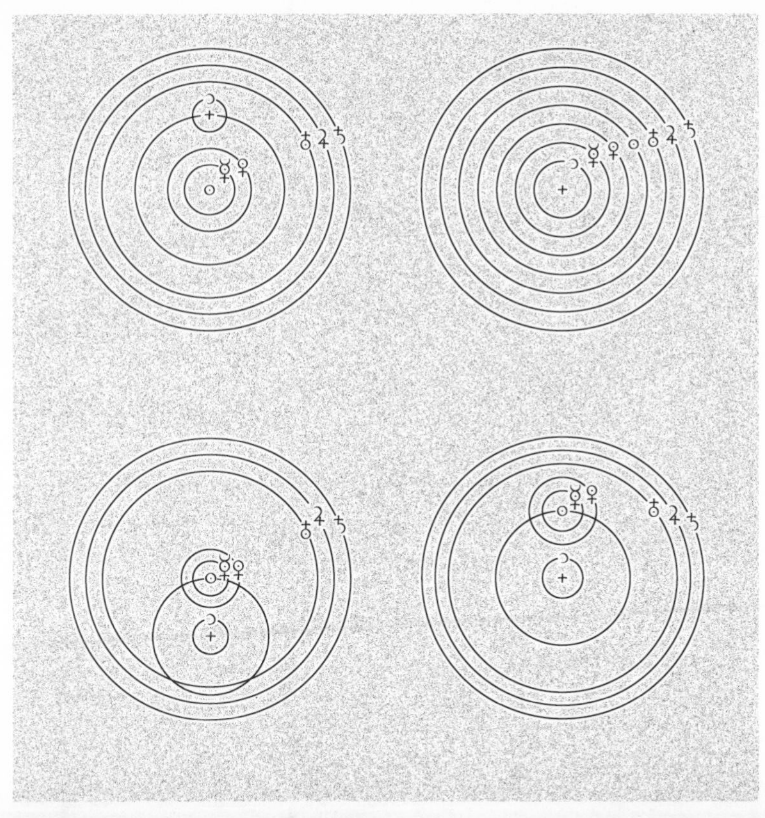

KEPLER'S VISION
ellipses and nested solids

Kepler noticed three things about planetary orbits: first, that they are ellipses (*so that a + b = constant; opposite, bottom*) with the Sun at one focus; second, that the area of space swept out by a planet in a given time is constant; third, that the period T of a planet relates to R, its semimajor axis (average orbit), so that T^2/R^3 is a constant throughout the entire solar system.

Looking for a geometric or musical solution to the orbits, Kepler observed that six heliocentric planets meant five intervals. The famous geometric solution he tried was to fit the five Platonic solids between their spheres (*opposite, top and detail below*).

In recent years, far from diminishing Kepler's vision, Einstein's laws actually showed that the tiny space-time effects caused by Mercury's faster (and therefore heavier and time-slowed) motion when nearer to the Sun affected the precessional rotation of the ellipses over thousands of years, thus reinforcing Kepler's shells.

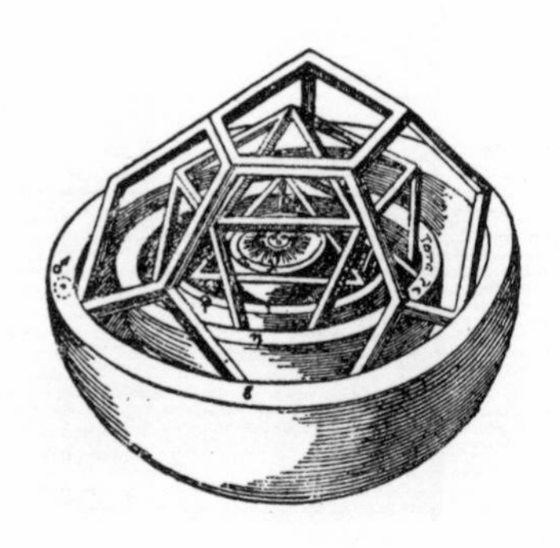

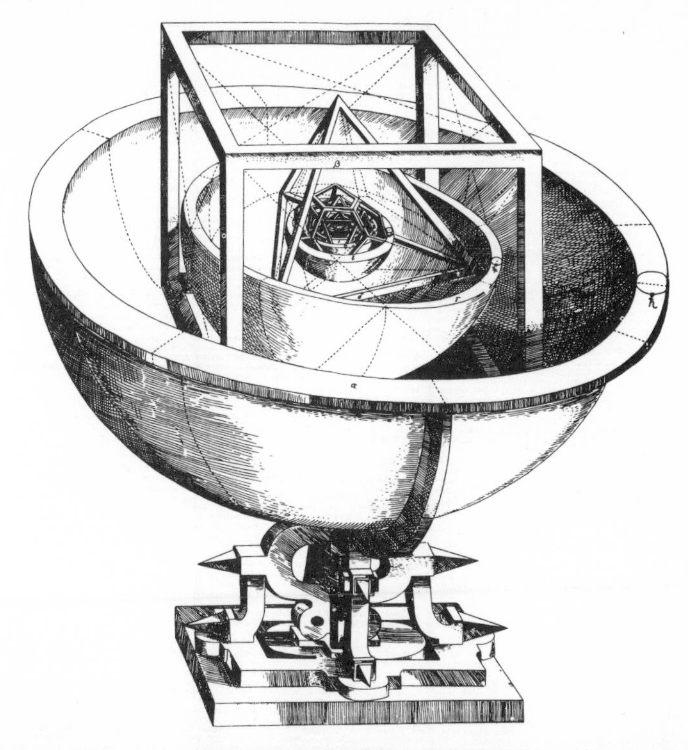
β
α

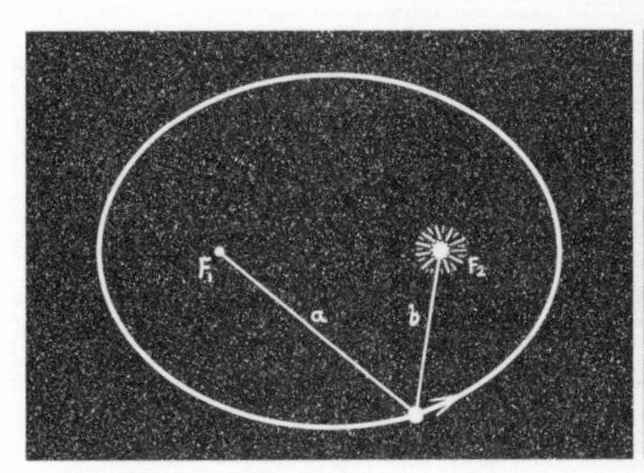
F_1
F_2
a
b

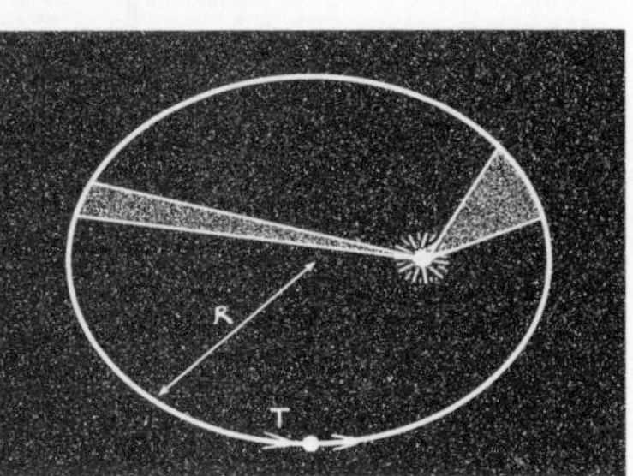
R
T

THE MUSIC OF THE SPHERES

planets playing in tune

In ancient times the seven musical notes were assigned to the seven heavenly bodies in various symbolic arrangements (*opposite, top*). With his accurate data, Kepler now set about precisely calculating these long imagined "Harmoniae Mundi." He particularly noticed that the ratios between planets' extreme angular velocities were all harmonic intervals (*opposite, center; after Jocelyn Godwin*). More recently, in 1968, research by A. M. Molchanov revealed the entire solar system as a "tuned" quantum resonant structure, with Jupiter as the conductor.

Music and geometry are very close bedfellows, and Carl von Weizsacker's 1948 particle-cloud formation theory of the condensation of the planets (*opposite, bottom; after Murchie and Warshall*) throws yet more dappled light on these elusive orbits. It might appear fanciful, were it not for the fact that two nested pentagons define Mercury's orbital shell (99.4%), the empty space between Mercury and Venus (99.2%), Earth's and Mars's relative mean orbits (99.7%), and the space between Mars and Ceres (99.8%). Three nested pentagons define the space between Venus and Mars (99.6%) or Ceres' and Jupiter's mean orbits (99.6%). A hidden pattern?

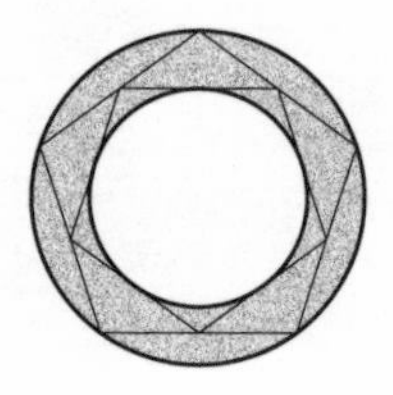

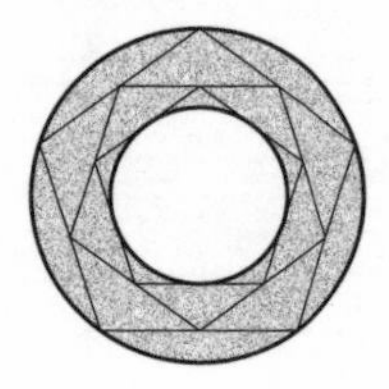

ANCIENT EGYPTIAN SYSTEM

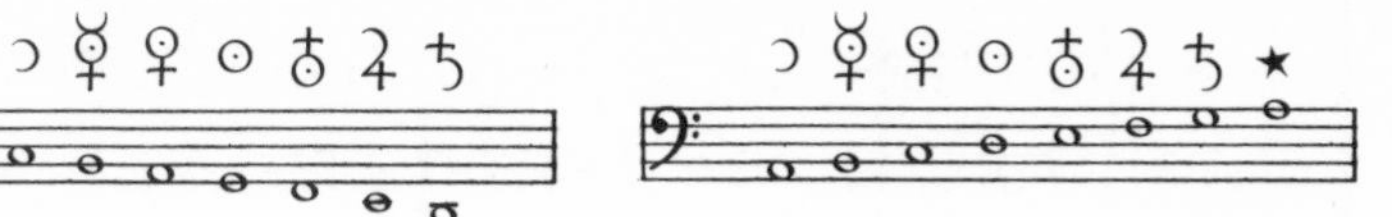

CICERO - SCIPIO'S DREAM

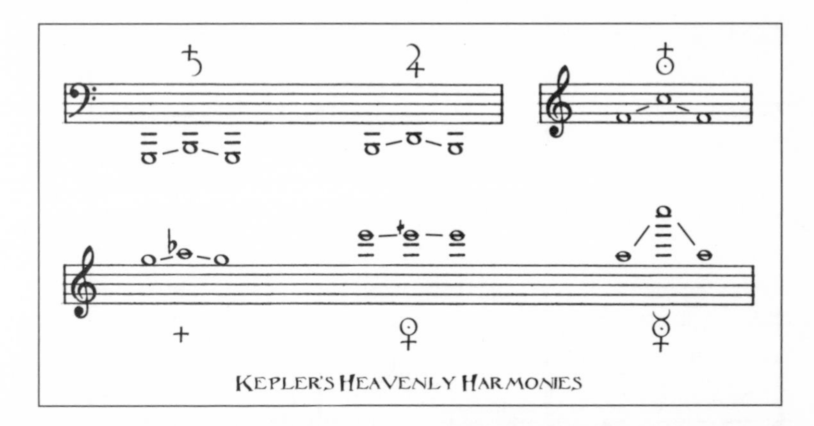

KEPLER'S HEAVENLY HARMONIES

Bode's Law and Synods

harmonics and rhythmic kisses

There have been numerous attempts to discover a simple pattern in the orbits and periods of the planets. A basic logarithmic graph shows clear underlying order in the planetary orbits (*opposite, top; after Ovendon and Roy*), and this also exists in various moon systems.

In 1778 J. Bode codified an observation made by J. Titius in 1766: If 4 is added to the series (0), 3, 6, 12, 24, 48, 96, 192 and 384, giving 4, 7, 10, 16, 28, 52, 100, 196 and 388, respectively, the new numbers, with Earth at 10 units, fit the planetary orbital radii quite well except that they leave out Neptune. In particular, the formula predicted a missing planet at twenty-eight units between Mars and Jupiter, and on January 1, 1801 Giuseppe Piazzi discovered Ceres, the largest of the asteroids in the asteroid belt, in the correct orbit.

Planets' orbital periods sometimes occur as simple ratios of each other, a famous example being the 2:5 ratio of Jupiter and Saturn (99.3%). Uranus, Neptune, and tiny Pluto are especially rhythmic and harmonic, displaying a 1:2:3 ratio of periods, Uranus's and Neptune's adding to produce Pluto's (99.8%).

Like a whirlpool, inner planets orbit the Sun much faster than outer planets and the table (*opposite, bottom*) shows the number of days between two planets' kisses, passes, or near approaches, properly called *synods*. Does Earth experience any harmonics? Well, we have two planetary neighbors—Venus sunside and Mars spaceside—and the figures reveal that we kiss Mars three times for every four Venus kisses (99.8%). So an ultraslow 3:4 rhythm or a deep musical fourth is being played around us all the time.

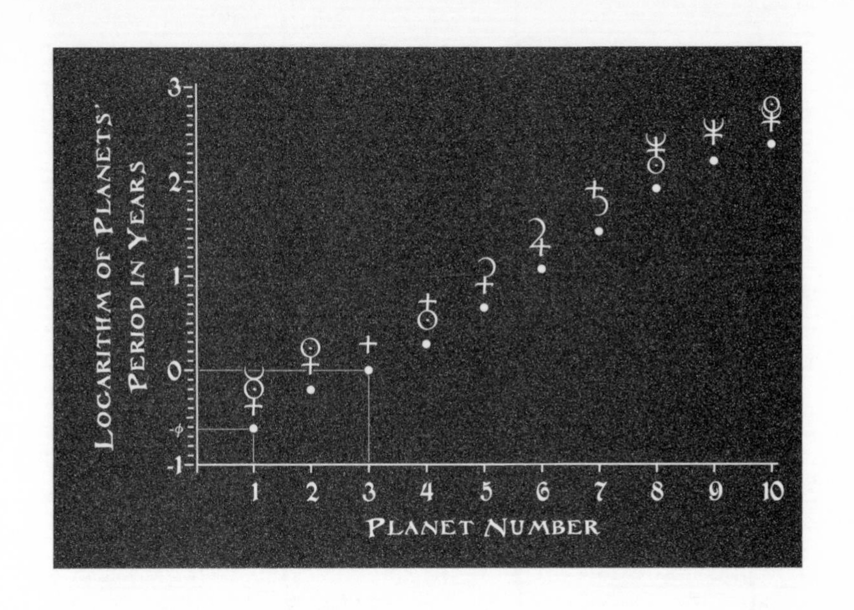

	☿	♀	+	♂	⚳	♃	♄	♅	♆	♇
☿	∞	144.6	115.9	100.9	92.83	89.79	88.70	88.22	88.10	88.05
♀	144.6	∞	583.9	333.9	259.4	237.0	229.5	226.4	225.5	225.3
+	115.9	583.9	∞	779.9	466.7	398.9	378.1	369.7	367.5	366.7
♂	100.9	333.9	779.9	∞	1,162	816.5	733.9	702.7	694.9	692.2
⚳	92.83	259.4	466.7	1,162	∞	2,744	1,991	1,777	1,728	1,712
♃	89.79	237.0	398.9	816.5	2,744	∞	7,252	5,045	4,669	4,551
♄	88.70	229.5	378.1	733.9	1,991	7,252	∞	16,570	13,100	12,210
♅	88.22	226.4	369.7	702.7	1,777	5,045	16,569	∞	62,890	46,440
♆	88.10	225.5	367.5	694.9	1,728	4,669	13,100	62,890	∞	179,800
♇	88.05	225.3	366.7	692.2	1,712	4,551	12,210	46,440	179,800	∞

THE NUMBER OF DAYS BETWEEN TWO PLANETS' KISSES

The Inner Planets
Mercury, Venus, Earth, and Mars

The solar system is divided by an asteroid belt into two halves. In the inner region four small, rocky planets orbit the Sun, in the outer region four huge gas and ice planets slowly trundle around.

The Sun has still not given up its secrets. Mostly hydrogen and helium, and an element factory, it is also a giant fluid geometric magnet, 15 million°C at its core, 6,000°C at the surface. It blows a particle wind through the entire solar system and its sunspots and huge solar flares affect electronics on Earth.

Mercury is the first planet. Mostly solid iron, it is a cratered, atmosphereless world, 400°C in the sunshine, -170°C in the shade.

Venus, the second planet, is a cloud-shrouded greenhouse world. On the surface the temperature is a staggering 480°C, and the carbon-dioxide-rich atmosphere is ninety times denser than Earth's. An apple here would be instantly incinerated by the heat, crushed by the atmosphere, and finally dissolved in sulphuric acid rain.

Earth is the third planet, the one with the miracle of life and just one large Moon.

Mars, the fourth planet, is a rocky red world, just above freezing. Ice caps cover the poles under a thin atmosphere. Riverbeds suggest that Mars may once have had oceans but they are long gone now, and today dust storms regularly envelop the planet for days. Huge dead volcanoes, one three times larger than Mount Everest, stand witness to a bygone age. Mars has two tiny moons.

Beyond Mars is the asteroid belt, dominated by Ceres, and beyond that is the realm of the giants.

SIZES OF THE INNER PLANETS

TILTS AND ECCENTRICITIES OF THE ORBITS

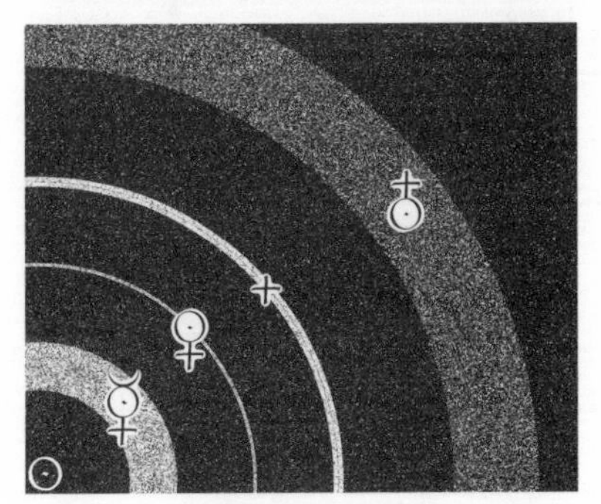

SUN CENTERED

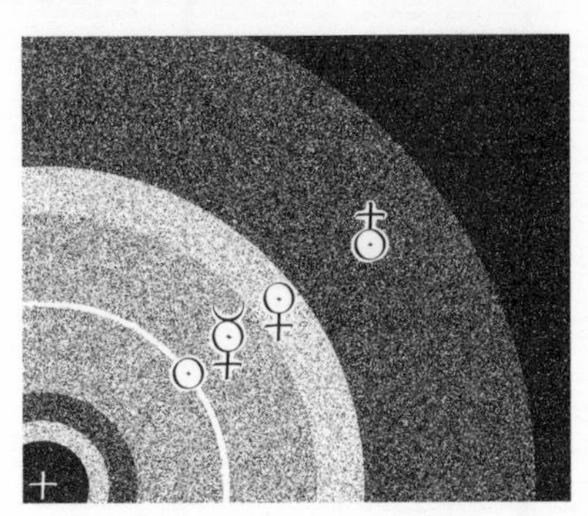

EARTH CENTERED

The Orbits of Mercury and Venus

a very simple aide-memoire

There are few things simpler than a circle. Despite Kepler's ellipses, and Newton and Einstein setting them spinning, the planetary orbits can still be thought of as orbital "shells" or sets of concentric circles, centered on the Sun, with the eccentricity of the ellipse thickening the circle slightly, or giving the spheres a shell (*see Kepler's diagram, page 13*).

One of the first things you can do with circles is to put three of them together so that they all touch. Amazingly, the orbits of the first two planets of the solar system are hiding in this simple design. If Mercury's mean orbit passes through the centers of the three circles then Venus's encloses the figure (99.9%).

This is a simple trick to remember—you see it all around you all the time, in the home, in design, art, architecture, and nature. Every time you pick up three glasses or push three balls together you create the first two planets' circular orbits (the ideal or orbit purely defined by their periods). There must be a reason for this beautiful fit between the ideal and the manifest; maybe some bright twenty-first-century scientist will find it. Until then, it remains a beautiful coincidence.

Venus (or Aphrodite or Freya) is the traditional goddess of love, harmony, and beauty, and has the most perfectly circular orbit of all the planets in the solar system. Mercury (or Hermes, Thoth, or Wotan) is the ancient god of geometry, communication, and initiation. Mercury has a highly elliptical orbit (*its inner and outer distances from the Sun are shown by the dotted line opposite, top*).

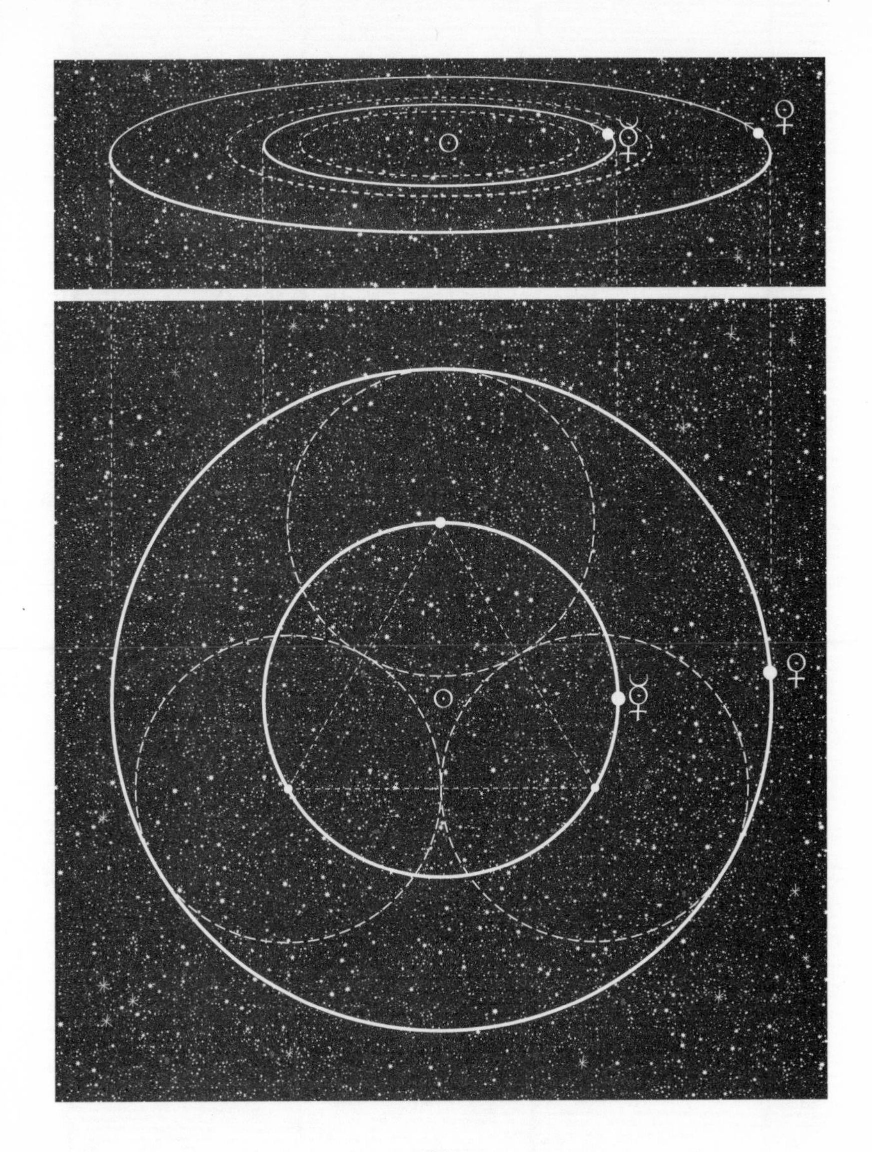

Making Sense of the Pictures

tips on things that appear in this book

Seen from Earth, day or night, the Sun appears to move slowly to the left against the stars (to the right in the southern hemisphere), taking a year to return to the same star. The Moon swiftly circles around in the same direction every month, taking 27.3 days to return to a star, or 29.5 days to catch up with the Sun. Venus and Mercury oscillate around the Sun, coming and going, as the Sun itself slowly trundles around its yearly circle. Imagine standing on Venus. The Sun moves faster against the stars, and Mercury is closer, whirling around the Sun like a couple waltzing.

Every pair of planets creates a single dance. It doesn't matter which of the two you stand on, your partner's dance around you will be the same. It is a shared experience. Mercury's evolving waltzes with Earth and Venus are shown (*opposite, top*). Earth and Mercury roughly kiss 22 times in 7 years, though the ancient Greeks also knew of a more accurate 46 year, 145 synod cycle. Mercury and Venus are beautifully in tune after just 14 kisses.

In the pages that follow we will meet the Golden Ratio, ϕ or *phi*. It is found throughout every pentagram and in the Fibonacci series of numbers, where the ratio between successive pairs of numbers gives closer and closer approximations to it (*opposite, bottom*). The Golden Ratio is essentially 0.618, but since 1 divided by it is 1.618 (which is the same as adding 1 to it), and 1.618 times 1.618 equals 2.618 (the same as adding 1 more), it can appear as any of these values. The Golden Ratio is found throughout organic life and was widely used in ancient art and architecture.

240 DAYS
770 DAYS
2,030 DAYS
MERCURY AND VENUS'S DANCE
470 DAYS
1,390 DAYS
2,510 DAYS
MERCURY AND EARTH'S DANCE

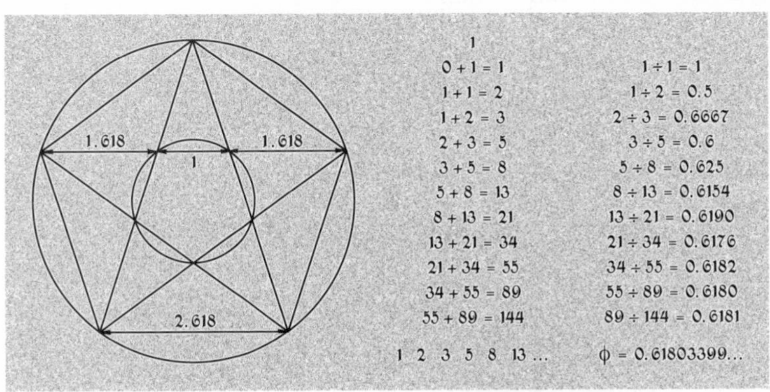
1.618
1.618
1
2.618
1
0 + 1 = 1
1 + 1 = 2
1 + 2 = 3
2 + 3 = 5
3 + 5 = 8
5 + 8 = 13
8 + 13 = 21
13 + 21 = 34
21 + 34 = 55
34 + 55 = 89
55 + 89 = 144
1 2 3 5 8 13 ...
1 ÷ 1 = 1
1 ÷ 2 = 0.5
2 ÷ 3 = 0.6667
3 ÷ 5 = 0.6
5 ÷ 8 = 0.625
8 ÷ 13 = 0.6154
13 ÷ 21 = 0.6190
21 ÷ 34 = 0.6176
34 ÷ 55 = 0.6182
55 ÷ 89 = 0.6180
89 ÷ 144 = 0.6181
φ = 0.61803399...

The Kiss of Venus

our most beautiful relationship

Other than the Sun and Moon, the brightest point in the sky is Venus, morning and evening star. It is our closest neighbor, kissing us every 584 days as it passes between us and the Sun. Each time one of these kisses occurs, the Sun, Venus, and the Earth line up two-fifths of a circle farther around the starry zodiacal circle—so a pentagram of conjunctions is drawn. Seen from Earth the Sun moves around the zodiac while Venus whirls around the Sun, drawing an astonishing pattern over exactly eight years (99.9%) [or thirteen Venusian years (99.9%)]. Small loops are made when Venus in her dazzling kiss seems briefly to reverse direction against the background stars (*shown below as seen from Earth*). Notice the Fibonacci numbers we have just met, 5, 8, and 13. The periods of Earth and Venus are also closely related as $\phi:1$ (99.6%).

The fivefold nature of Venus and Earth's dance extends to their closest and farthest distances from each other. At the bottom of the opposite page we see Venus's *perigee* and *apogee* defined by two pentagrams. The body of space one draws around the other is thus sized $1:\phi^4$ (99.9%).

All these diagrams also apply to Venus's experience of Earth.

THE PERFECT BEAUTY OF VENUS
the things they don't teach you at school

With the Sun back in the center, let us again look at the orbits of Venus and the Earth and draw a line between the two planets positions every couple of days (*below left*). Because Venus orbits faster it completes a whole circuit in the same time that the Earth completes just over a half-circuit (*below center*). If we keep watching for exactly eight years the pattern opposite emerges, the sun-centered version of the five-petaled flower on the previous page.

The ratio between Earth's outer orbit and Venus's inner orbit, i.e, their home, is intriguingly given by a square (*below right*) [99.9%].

Venus rotates extremely slowly on her own axis in the opposite direction to most rotations in the solar system. Her day is precisely two-thirds of an Earth year, a musical fifth. This exactly harmonizes with the dance opposite so that every time Venus and Earth kiss, Venus does so with her same face looking at the Earth. Over the eight Earth years of the five kisses, Venus will have spun on her own axis twelve times in thirteen of her years (*from Kollerstrom*). All exact and very beautiful. Mercury also displays a harmonious calendar as its day is two of its years, a musical octave.

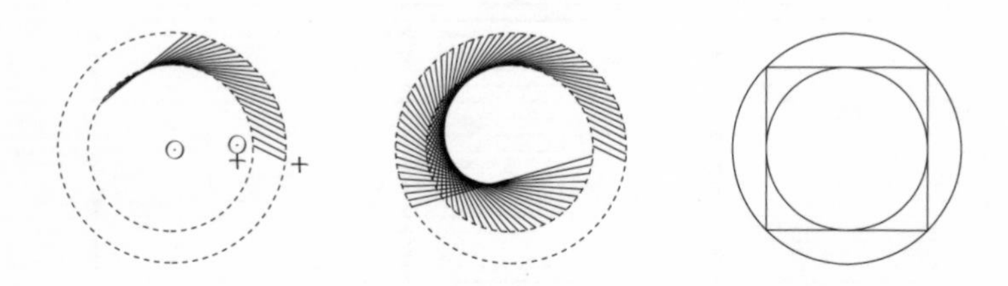

MERCURY AND EARTH

doubling up with more fives and eights

Can two planets' physical sizes double as a scaled picture of their orbits? It happens twice in the solar system, between Mercury and Earth, and between Saturn and Earth.

Mercury's and Earth's sizes and orbits are shown with coinciding "phive" and eightfold overlays (*top four images opposite*), all of which give the relative orbits and physical sizes of these two planets to around 99 percent accuracy, about the thickness of the pen line. All solutions are interchangeable at this accuracy, thus the octagram (*top right*) could also produce their relative orbits.

The diameter of Mercury's innermost orbit is suggested by the pentagram incircle (*center left*) (99.5%) and also happens to be the distance between the mean orbits of the two planets (99.7%).

It is interesting to expand on the three touching circles of page 21, which spaced Mercury and Venus. Center right we see the three circles again, and this time we add eight touching circles centered on Venus's orbit to produce Earth's mean orbit (99.99%)—the eight years of the five kisses perhaps?

Earth's and Saturn's relative orbits and sizes can be given by a fifteen-point star (*opposite, lower row*), which also famously gives the Earth's axial tilt.

Saturn and Earth incidentally have another more famous coincidental arrangement—Saturn takes the same number of years to go around the Sun as there are days between full Moons (99.8%). Lunacy? It is to the Moon that we now turn . . .

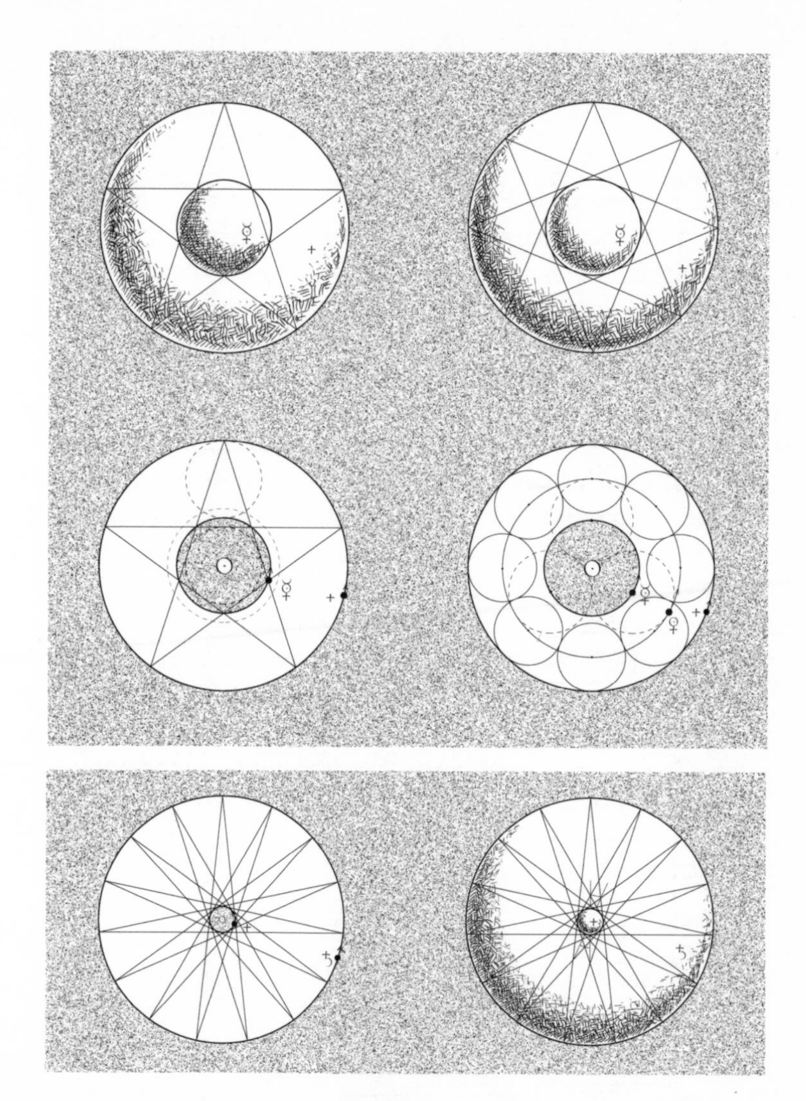

The Alchemical Wedding

three to eleven all around

From the surface of the Earth, the Sun and the Moon appear to be the same size. According to modern cosmology this is nothing but a coincidence. In ancient times the balance between these two primary bodies was seen as proof of the perfection of creation.

The size of the Moon compared to the Earth is three to eleven (99.9%). What this means is that if you pull the Moon down to the Earth, then the circle through the center of the heavenly Moon will have a circumference equal to the perimeter of a square enclosing the Earth. The ancients seem to have known about this, and hidden it in the definition of the mile, which seems to have been chosen with extraordinary care (*opposite, bottom; after John Michell & Dan Ward*).

The Earth-Moon proportion is also precisely invoked by our two planetary neighbors, Venus and Mars (*Venus shown dancing around Mars below*). The closest: farthest distance ratio that each experiences of the other is, incredibly, 3:11 (99.9%). We orbit between them.

Quite by chance, 3:11 is 27.3 percent, and the Moon orbits the Earth every 27.3 days, also the average rotation period of a sunspot. The Sun and Moon do seem very much the unified couple.

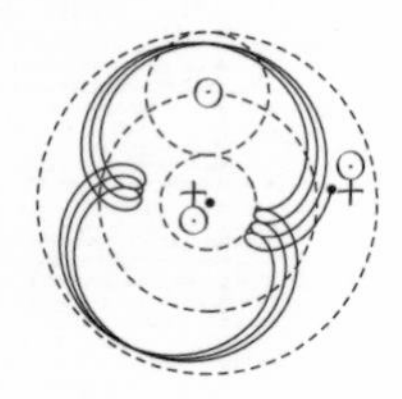

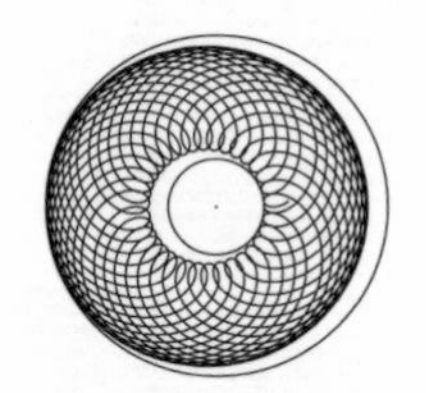

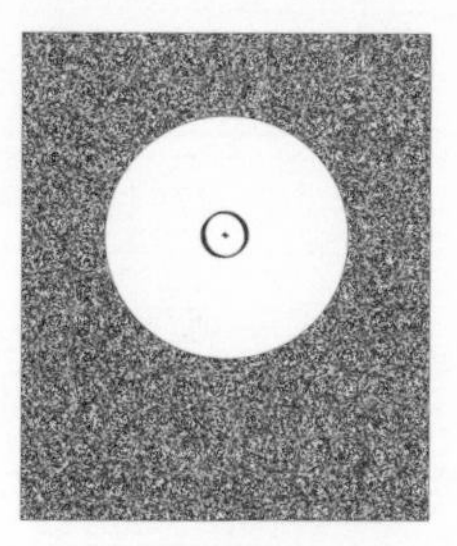

THE MOON, A TOTAL SOLAR ECLIPSE, AND THE SUN, AS SEEN FROM EARTH

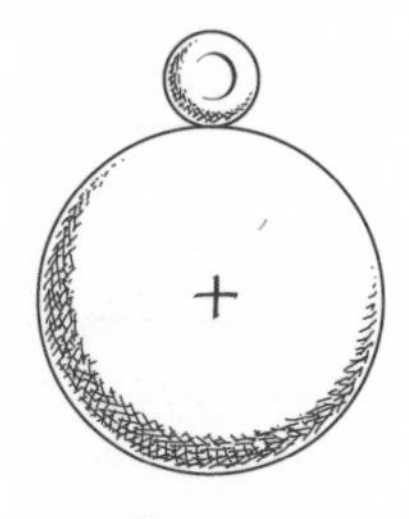
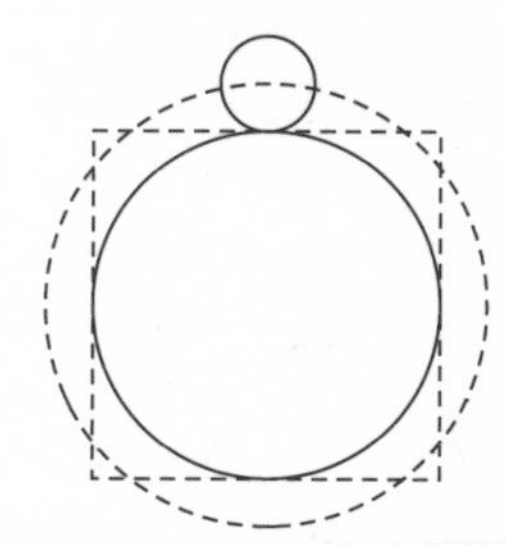
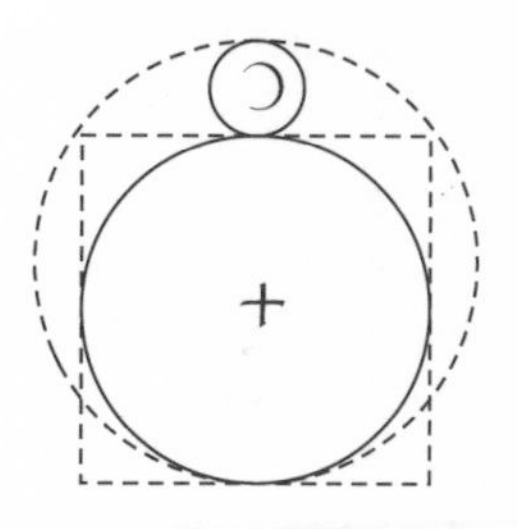

THE SIZES OF THE MOON AND THE EARTH "SQUARE THE CIRCLE"
THE DOTTED-LINE SQUARE AND CIRCLE ARE THE SAME LENGTH OF STRING

MILES OF MOON AND EARTH

RADIUS OF MOON = 1,080 MILES = 3 x 360

DIAMETER OF MOON = 2,160 MILES = 6 x 360 = 18 x 1 x 2 x 3 x 4 x 5

RADIUS OF EARTH = 3,960 MILES = 11 x 360 = 33 x 1 x 2 x 3 x 4 x 5

RADIUS OF EARTH + RADIUS OF MOON = 5,040 MILES
= 1 x 2 x 3 x 4 x 5 x 6 x 7 = 7 x 8 x 9 x 10

DIAMETER OF EARTH = 7,920 MILES = 8 x 9 x 10 x 11

THERE ARE 5,280 FEET IN A MILE
= (10 x 11 x 12 x 13) - (9 x 10 x 11 x 12)

Calendar Magic

the astonishing discoveries of Mr. Heath

Recent work by Robin Heath has revealed simple geometrical and mathematical tools that suggest order and form within the Sun-Moon-Earth system. Imagine we want to discover the number of full moons in a year (somewhere between twelve and thirteen). Draw a circle, diameter thirteen with a pentagram inside. Its arms will then measure 12.364, the number of full moons in a year (99.95%).

An even more accurate way of doing this is to draw the second Pythagorean triangle, which just happens to be made of numerals 5, 12, and 13 again, interestingly also the numbers of the keyboard, and of Venus (*page 26*). Dividing the 5 side into its harmonic 2:3 gives a new length 12.369, the number of full moons in a year (99.999%)

The Moon seems to beckon us to look further. We all know that six circles fit around one on a flat surface (giving the numbers six and seven). Twelve spheres pack perfectly around one in our familiar three-dimensional space (our familiar twelve and thirteen). We seem to be moving up in sixes. Could *eighteen* time-spheres fit around one in a fourth dimension of time to give the numbers eighteen and nineteen? Incredibly, all of the current major time cycles of the Sun-Moon-Earth system can be expressed as simple combinations of the numbers 18, 19, and the Golden Section.

The Golden Section is evident in the pentagram, the icosahedron, the dodecahedron, and all living things. The orbits of the four inner planets all display its presence. Its values added to the magic number 18 produce 18, 18.618, 19, 19.618 and 20.618, which then multiply together as shown opposite. Coincidence or biophysics?

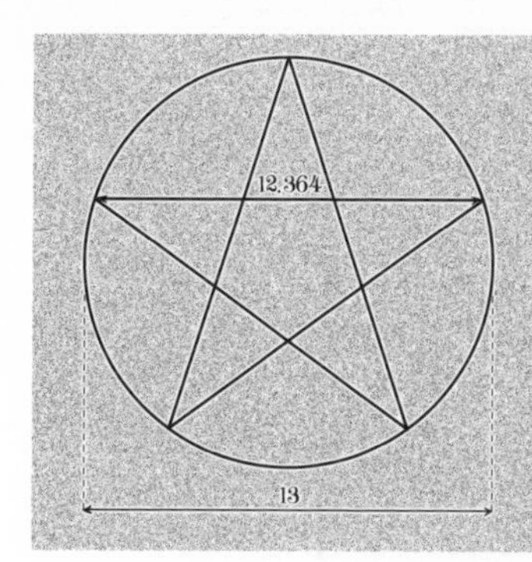

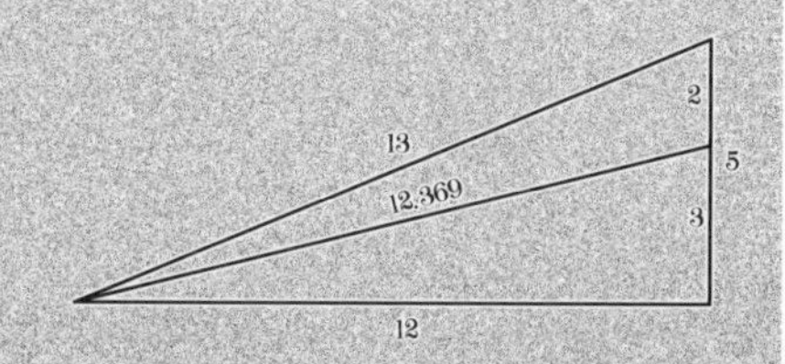

TWO ANCIENT TECHNIQUES FOR FINDING THE NUMBER OF FULL MOONS IN A YEAR

18 YEARS = THE SAROS ECLIPSE CYCLE (99.83%)

(SIMILAR ECLIPSES WILL OCCUR AFTER 18 YEARS)

18.618 YEARS = REVOLUTION OF THE MOONS NODES (99.99%)

(THE MOONS NODES ARE THE TWO PLACES WHERE THE SLIGHTLY OFFSET CIRCLES OF THE SUN AND MOONS ORBITS CROSS)

19 YEARS = THE METONIC CYCLE (99.99%)

(IF THERE IS A FULL MOON ON YOUR BIRTHDAY THIS YEAR - THERE WILL BE ANOTHER ONE ON YOUR BIRTHDAY IN 19 YEARS TIME)

THE ECLIPSE YEAR = 18.618 X 18.618 DAYS (99.99%)

(THE ECLIPSE YEAR IS THE TIME IT TAKES FOR THE SUN TO RETURN TO THE SAME ONE OF THE MOONS NODES. IT IS 18.618 DAYS SHORT OF A SOLAR YEAR (99.99%). THERE ARE 19 ECLIPSE YEARS IN A SAROS)

12 FULL MOONS = 18.618 X 19 DAYS (99.82%)

(12 FULL MOONS IS THE LUNAR OR ISLAMIC YEAR)

THE SOLAR YEAR = 18.618 X 19.618 DAYS (99.99%)

(THE SOLAR YEAR IS THE 365.242 DAY YEAR WE ARE USED TO)

13 FULL MOONS = 18.618 X 20.618 DAYS (99.99%)

(13 FULL MOONS IS ANOTHER 18.618 DAYS AFTER THE SOLAR YEAR)

Cosmic Football

Mars, Earth, and Venus spaced

The next planet out from Earth is the fourth planet, Mars. Kepler had tried a *dodecahedron* spacing the orbits of Mars and Earth and an *icosahedron* spacing Earth from Venus (*see the illustration on page 12*), and, coincidentally, it turns out he was very close to the mark.

The dodecahedron (made of twelve pentagons) and the icosahedron (made of twenty equilateral triangles) are the last two of the five perfect polyhedra (the *Platonic solids*). They form a dual pair, as each creates the other from the centers of its faces (*below*). Opposite, they appear in bubble form suspended inside Mars's spherical orbit. The dodecahedron magically produces Venus's orbit as the bubble within (*opposite, top*) [99.98 %], while the icosahedron defines Earth's orbit through its bubble centers (*opposite, bottom*) [99.9 %].

In the ancient sciences the icosahedron was associated with the element of water, so it is appropriate to see it emanating from our watery planet. The dodecahedron represented the fifth element of ether, the life force, here enveloping lively Earth, and perfectly defined by its two neighbors.

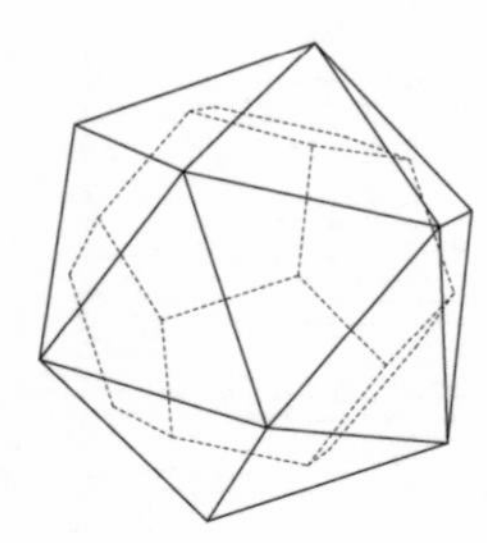

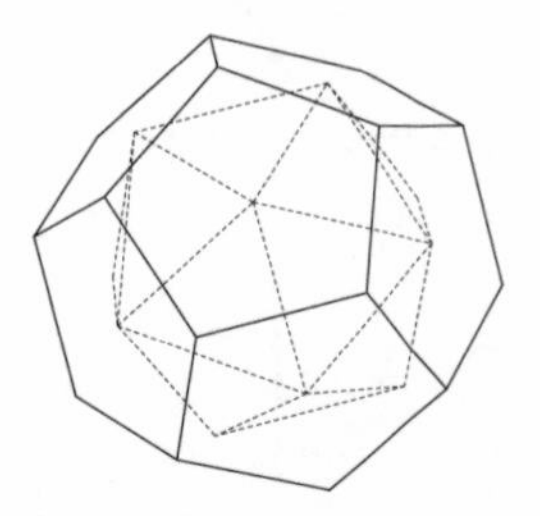

Venus and Mars
Earth and Mars

THE ASTEROID BELT

through the looking glass

We have reached the end of the inner solar system. Beyond Mars lies a huge space, the other side of which is the enormous planet Jupiter. It is in this space that the asteroid belt is found, thousands of large and small tumbling rocks, silicaceous, metallic, carbonaceous, and others. Like the gaps in Saturn's rings, there are spaces, termed *Kirkwood Gaps*, in the asteroid belt, cleared where orbital resonances with Jupiter occur. The largest gap is at the orbital distance that would correspond to a period of one-third that of Jupiter.

By far, the largest of the asteroids is Ceres, comprising over one-third of the total mass of all of them. Ceres is about the size of the British Isles and produces a perfect eighteenfold pattern with Earth (*see page 57, top left*).

Bode's Law predicted something at the distance of the asteroid belt (*see page 16*), but it was Alex Geddes who recently discovered the extraordinary mathematical relationship between the four small inner planets and the four outer gas giants. Their orbital radii magically "reflect" around the asteroid belt and multiply as shown below and as illustrated opposite to produce two enigmatic constants. Yet again we find ourselves peering at a simple pattern we cannot explain.

$Ve \times Ur = 1.204\ Me \times Ne$

$Me \times Ne = 1.208\ Ea \times Sa$

$Ea \times Sa = 1.206\ Ma \times Ju$

$Ve \times Ma = 2.872\ Me \times Ea$

$Sa \times Ne = 2.876\ Ju \times Ur$

$(Ve \times Ma \times Ju \times Ur = Me \times Ea \times Sa \times Ne)$

The asteroid belt is unlikely to be the remains of a small planet as no sizeable body could ever have formed so close to Jupiter.

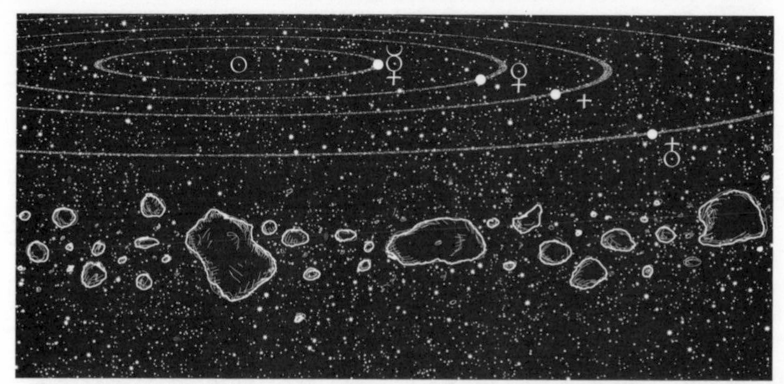

BETWEEN THE INNER AND OUTER REGIONS OF THE SOLAR SYSTEM

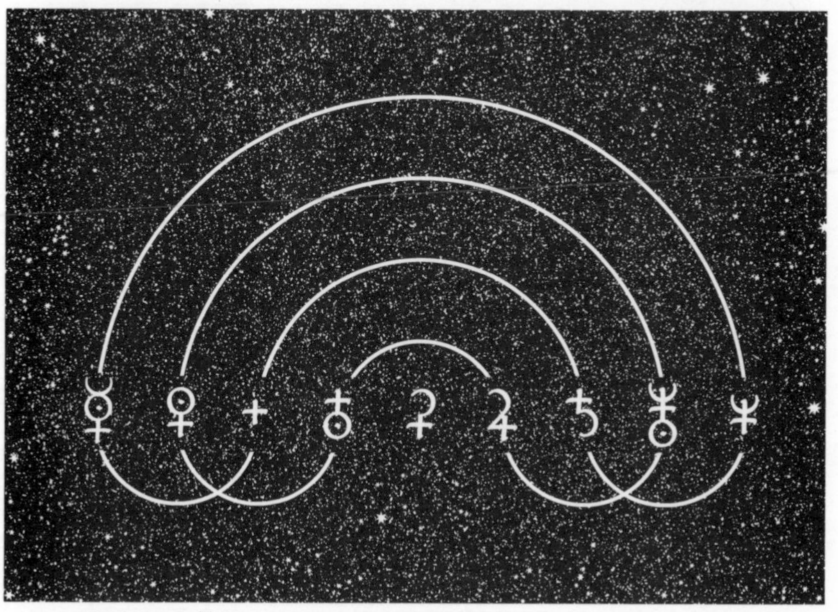

CERES AND THE ASTEROID BELT AS THE MULTIPLICATION MIRROR

THE OUTER PLANETS

Jupiter, Saturn, Uranus, Neptune, and beyond

Beyond the asteroid belt we come to the realm of the gas and ice giants, Jupiter, Saturn, Uranus, and Neptune.

Jupiter is the largest planet, and its magnetic field is the largest object in the solar system. Ninety percent hydrogen, it is nevertheless built around a rocky core like all the giant planets. Metallic and liquid hydrogen surrounds this core. The famous Red Spot is a storm, larger than Earth, that has raged now for hundreds of years. Jupiter's moons are numerous and fascinating: One of the four largest, Io, is the most volcanic body in the solar system; another, Europa, may have warm oceans of water beneath its icy surface.

Saturn, with its beautiful system of rings, is the second largest planet. Its structure beneath its clouds is much the same hydrogen and helium mix as Jupiter. A large number of moons have been discovered, the largest of which is Titan, a world the size of Mercury with all the building blocks for life except warmth.

Beyond Saturn is Uranus, which orbits on its side. Winds gust on the equator at six thousand times the speed of sound.

Next is Neptune, like Uranus an ice world of water, ammonia, and methane. The largest moon, Triton, has nitrogen ice caps and spews geysers of liquid nitrogen high into the atmosphere.

Finally comes the tiny planet Pluto with its large moon Charon, and beyond, the primordial swarm of the Kuiper Belt, from where Pluto probably came. Eventually, stretching a third of the way to the nearest star, is the sphere of icy debris of the Oort Cloud, home to the comets, which occasionally fall toward the Sun and inner planets.

SIZES OF THE OUTER PLANETS

TILTS AND ECCENTRICITIES OF THE ORBITS

SUN CENTERED

EARTH CENTERED

FOURS

Mars, Jupiter, and massive moons

An asteroid belt and 550 million kilometers separate Mars's and Jupiter's orbits, a greater distance than Earth's orbit. Jupiter is the first and largest of the gas giants, the vacuum cleaner of the solar system. If Jupiter had gathered only slightly more material during its long and ongoing formation its internal pressures would have turned it into a star and we would have had a second Sun.

The top diagram opposite shows a simple way to draw the orbits of Mars and Jupiter from four touching circles or a square (99.98%). It is a proportion commonly seen in church windows and railway stations. Shown below on this page is a pattern from the same family, which spaces Earth's and Mars's mean orbits (99.9%).

Jupiter has four particularly large moons, discovered by Galileo. The two largest, Ganymede and Callisto, are the size of the planet Mercury and produce one of the most perfect space-time patterns in the solar system. An observer living on either moon would experience the motions of the other in space and time as the beautiful harmonic fourfold diagram shown opposite.

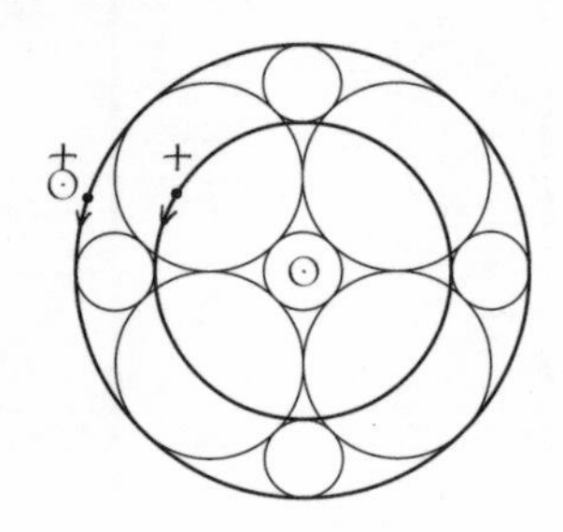

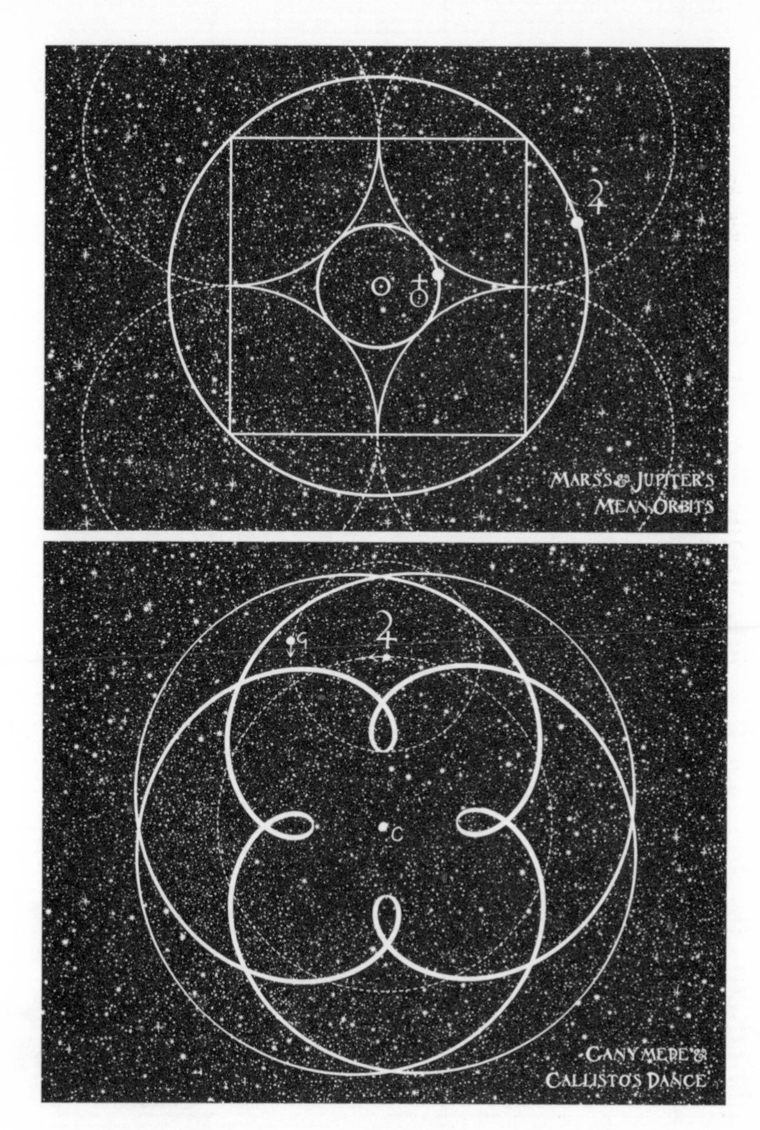
MARS'S & JUPITER'S
MEAN ORBITS
GANYMEDE &
CALLISTO'S DANCE

OUTER MOONS
harmonic patterns

Four groups of moons orbit Jupiter. The first two groups each have four moons and look very like a small model of the whole solar system—four small inner bodies followed by four giants.

The second group of four large moons, the *Galileans*, is itself divided into two rocky worlds, Io and Europa, then two massive gas-and-ice moons the size of planets, Ganymede and Callisto. The diagram below shows just how big some of these moons are.

The grouping into fours is very striking. Each of the four groups has its own general moon size, orbital plane, period, and distance from Jupiter (the inclinations of the four orbital planes of the four groups even add up to a quarter of a circle [99.9%]). Is there a reason why Jupiter displays such an affinity for the number four?

Saturn has over thirty moons, most shepherding and tuning the amazing rings with the larger bodies tending to be much farther out. Way beyond Saturn's rings are three moons—the gigantic Titan, tiny Hyperion, and, farther out still, Iapetus.

The picture opposite shows further harmonic patterns: two more from Jupiter's largest moons, two experienced by Saturn's outer moons, and two from the outer planets of the solar system.

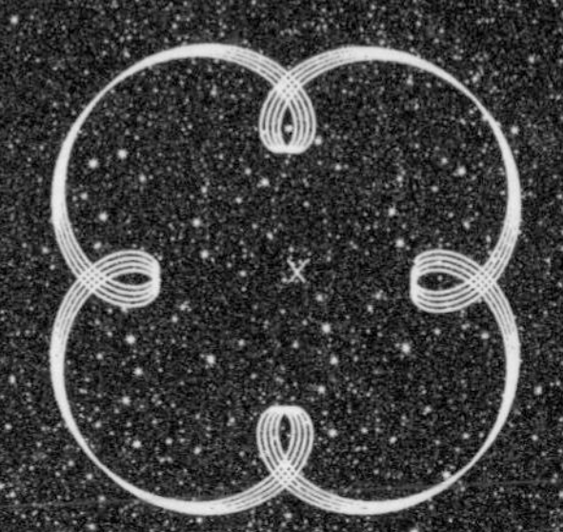

TITAN & IAPETUS

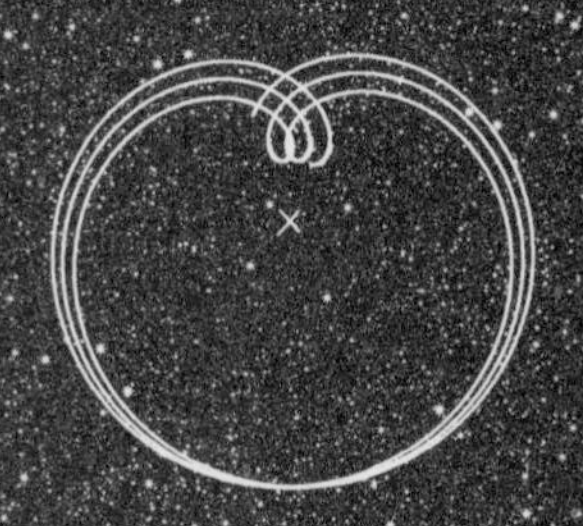

Jupiter's Giant Seal

huge hexagrams and affirmatory asteroids

Jupiter, the largest planet, was once king of the ancient gods. A delightful feature of its orbit is its pair of asteroid clusters, the Trojans, which move around Jupiter's orbit, 60° ahead of it and 60° behind (*opposite*). This partnership perpetually rotates around the Sun as though held in place by the spokes of a wheel. The positions of the Trojan clusters are known as the *Laplace Points*, with Sun, Jupiter, and Trojans forming gravitationally balanced equilateral triangles.

Just for the fun of it, if we now join the spokes as shown opposite then three hexagrams can be seen to produce Earth's mean orbit from Jupiter's (99.8%)—a very easy trick to remember. Earth's and Jupiter's orbits are thus lurking in every crystal. Another name for such a six-point star is a *Star of David* or *Seal of Solomon*, a kingly design indeed, by Jove! Exactly the same proportion may be created by spherically nesting three cubes, three octahedra, or any threefold combination of them inside Jupiter's orbit (*two possibilities are shown below*); the tiny sphere in the middle is Earth's orbit. All pure coincidence!

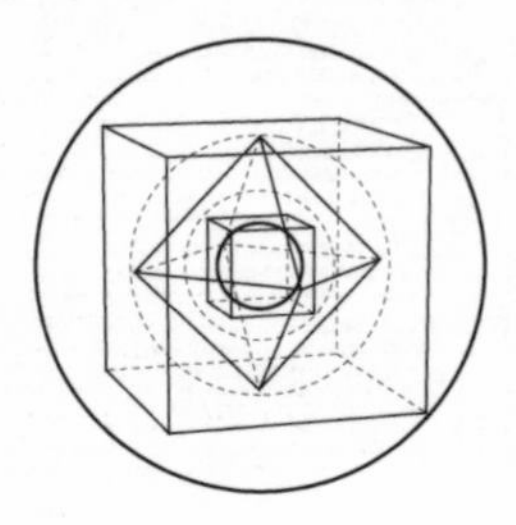

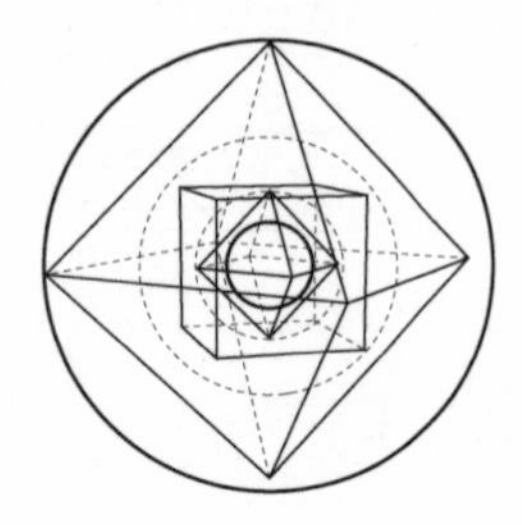

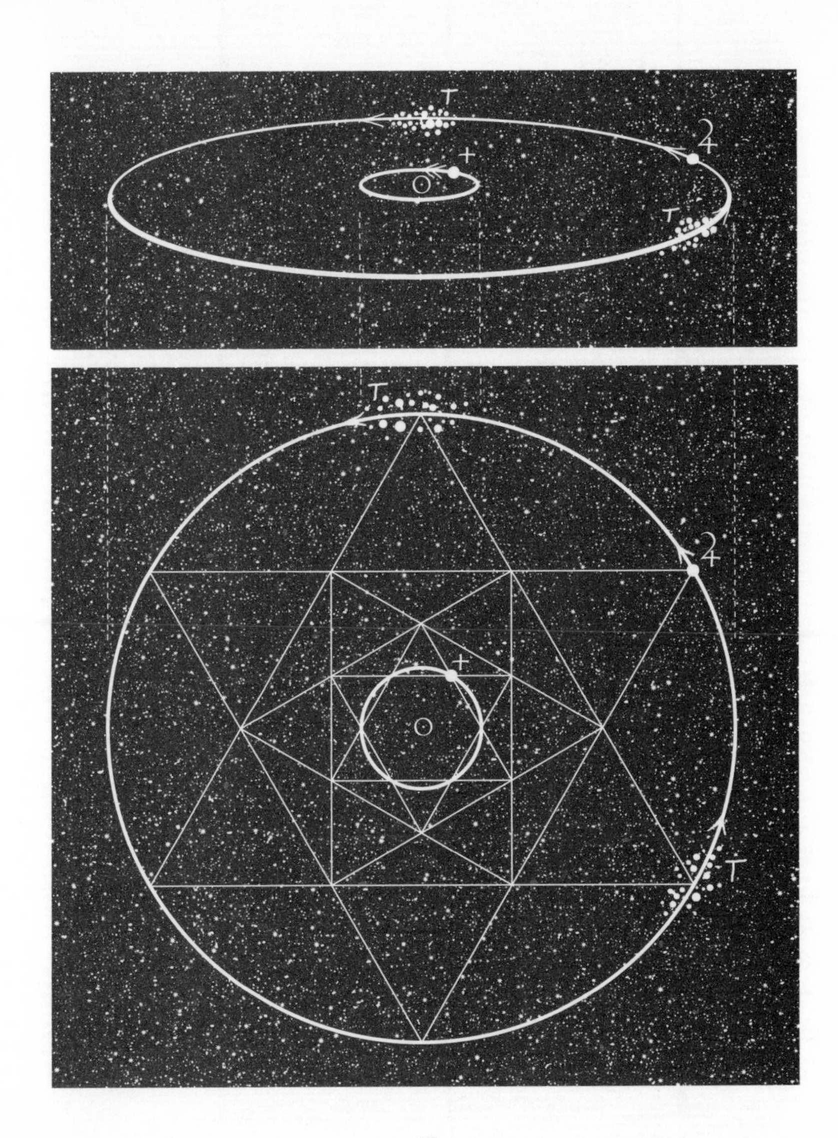
T
♃
T
T
♃
T

THE GOLDEN CLOCK

Jupiter and Saturn seen from Earth

Jupiter and Saturn are the two largest planets of the solar system and ruled the outer two spheres of the ancient system. In ancient mythology, Saturn was Chronos, the Lord of Time.

The top two diagrams opposite show the close 5:2 ratio of their periods. Top left we see their dance, viewing Jupiter from Saturn (Saturn's experience of Jupiter is the same). The beautiful threefold harmonic is immediately apparent, spinning slowly because of the slight miss in the harmony. From Earth, this pattern is seen as an important triangular sequence of conjunctions and oppositions of Jupiter and Saturn, who come together every twenty years. Top right we see the hexagram created by these positions—with conjunctions marked on the outside of the zodiac and oppositions marked inside. This is a diagram that has been known for thousands of years. The planets move counterclockwise around the circle of the ecliptic, shown here as a dotted line, starting at twelve o'clock, Jupiter moving faster than Saturn.

The bottom diagram shows the relative speeds of orbit of Earth, Jupiter, and Saturn. We start with the three planets in a line at twelve o'clock. Earth orbits much faster than the outer planets and makes a whole annual circuit of the Sun (365.2 days) before lining up with slow Saturn again for a synod after 378.1 days. Three weeks later it lines up with Jupiter (after 398.9 days).

Richard Heath recently revealed that the Golden Ratio is defined here in time and space to an accuracy of 99.99 percent! It should come as no surprise to discover the two giants of our solar system reinforcing the proportion of life on Earth.

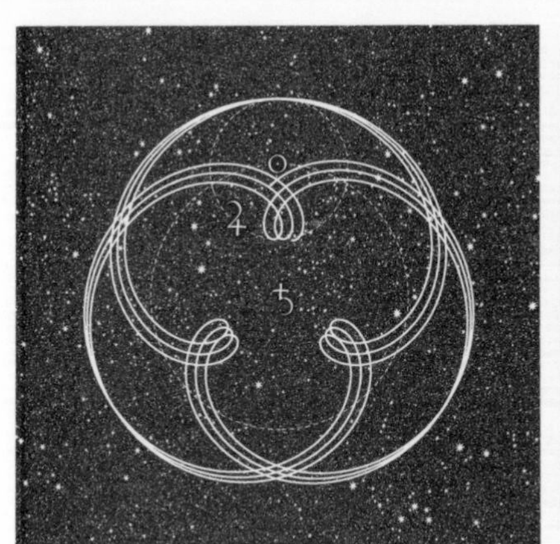

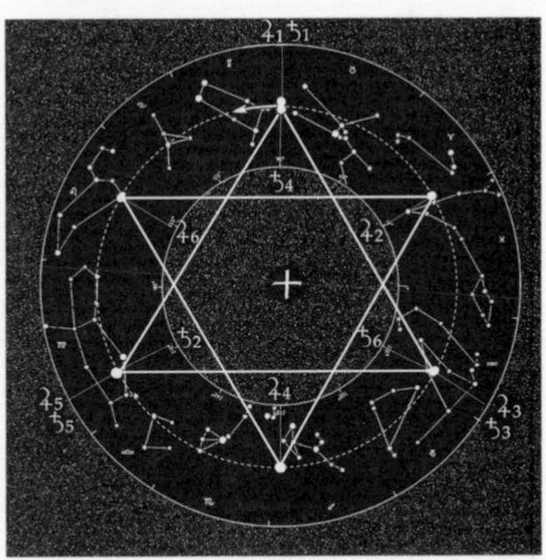

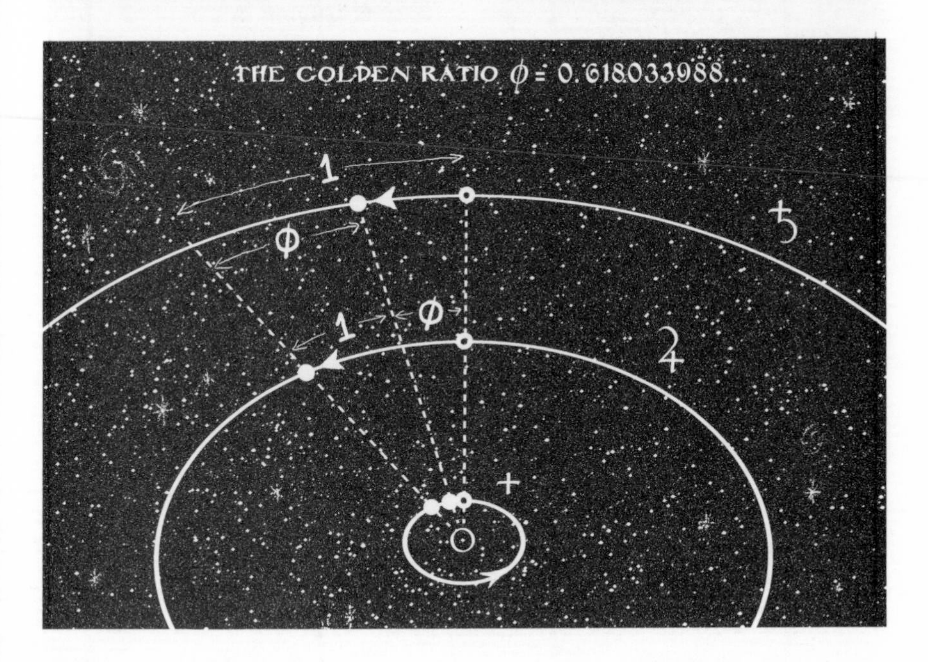
THE GOLDEN RATIO ϕ = 0.618033988...
1
ϕ
1
ϕ

Octaves out There

threes and eights again

Uranus was discovered in 1781 by William Herschel; it is the third largest planet in the solar system, orbiting on its side with twenty-one known moons and a faint system of rings. The diagram opposite shows a simple and memorable method for proportioning the outer, mean, and inner orbits of Uranus, Saturn, and Jupiter using an equilateral triangle and an octagram. The next time you see either of these shapes, just remember what you are looking at.

One way of depicting the musical *octave* (a halving or doubling of frequency or wavelength) is by an equilateral triangle, as the inscribed circle has a diameter half that of the containing circle.

Jupiter's and Saturn's orbits are in the proportion 6:11 (99.9%), the octave, or double, of the 3:11 Moon:Earth ratio (*see page 30*).

Saturn's orbit invokes π or *pi*—twice (*see below*). Its radius is the circumference of Mars's orbit (99.9%) and its circumference the diameter of Neptune's orbit (99.9%). You can now draw the solar system.

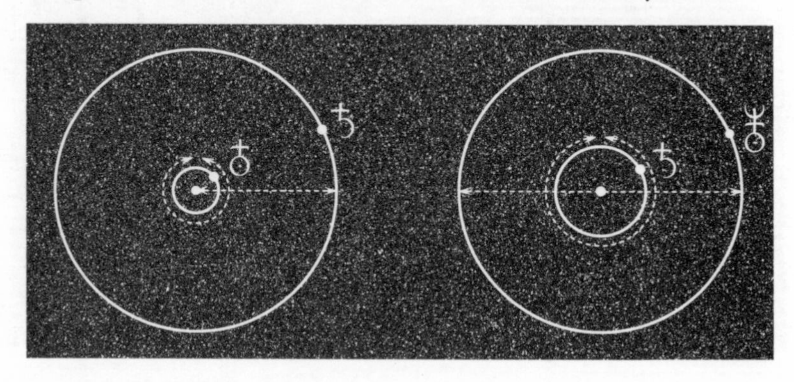

Harmonic Secrets

rings, shepherd moons, and five fives

Neptune was discovered in 1846. Like Uranus, it has a collection of moons and a faint ring system, thin etched circles hanging in space. The origins of planetary rings remain a mystery—the fine dust and tiny rocks may be the remains of moons destroyed in collisions, or they could be much older. Some gaps in Saturn's rings (*below*) are cleared by small "shepherd" moons, other spaces appear at Kirkwood distances, harmonic with one or more moons (*see page 36*).

Uranus's bright outermost ring has a diameter twice that of Uranus itself, an octave, and Neptune's innermost ring is two-thirds the size of its outermost (99.9%), a musical fifth (*opposite, top*). These harmonic proportions also mirror the local timing as Neptune's orbital period is twice that of Uranus, and Uranus's is two-thirds that of Pluto.

Uranus and Neptune dance around each other to create the beautiful shape shown opposite, which slowly spins so that every 4,300 years Neptune and Uranus both experience a perfect division of the zodiac into twenty-five kisses. Purely coincidentally, the tiny planet Chiron, orbiting between Saturn and Uranus, also measures out a perfect twenty-five around Uranus. These things happen.

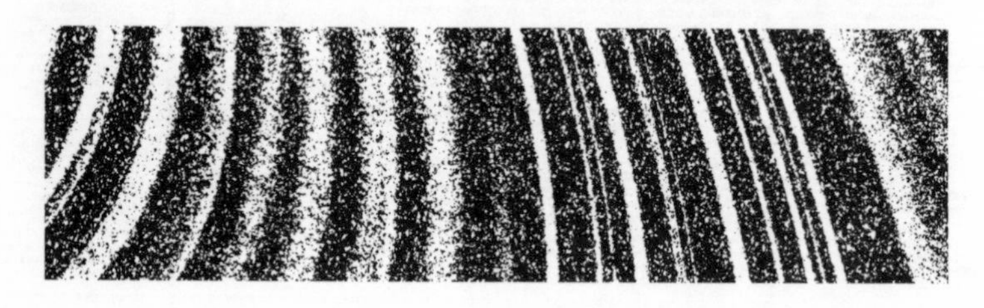

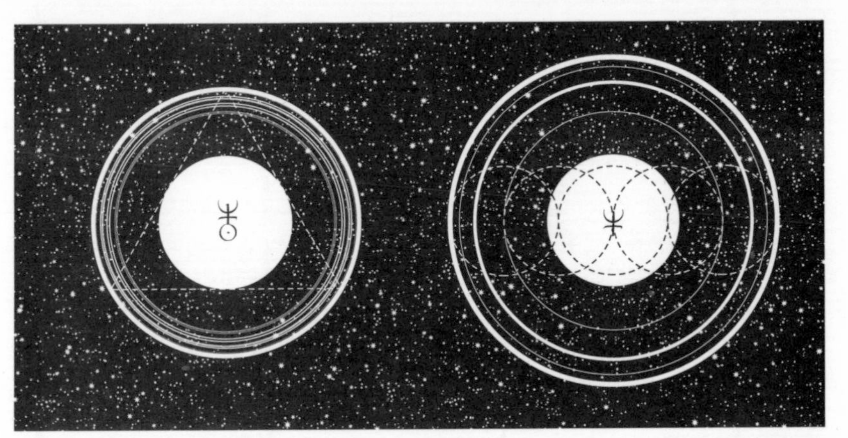

URANUS'S AND NEPTUNE'S RINGS

URANUS AND NEPTUNE'S DANCE

URANUS AND CHIRON'S DANCE

The Starry Signature
circumstantial evidence for life on Earth

Despite all the scientific discoveries over recent centuries we are possibly today as far from understanding what we are doing here as the ancients were from being able to build a pocket calculator. The ancients, however, pondered consciousness deeply, and held that life, or "the soul," was particularly akin to the arts of geometry and music. Through these arts they carefully investigated the relationship between "the One" and "the Few," for in music there are only so many notes in tune, and in geometry only so many shapes that fit. Kepler, Newton, Einstein, and many others to this day, have continued looking for simple and beautiful relationships in nature, expressing them as equations when they found them.

This book has shown simple and beautiful examples of harmony and geometry in the solar system. The Golden Ratio, long associated with life, and conspicuously absent from most modern equations, plays lovingly around Earth. Does this in some way have something to do with why we are here, and what we might be, and if so could these techniques be used to locate intelligent life in other solar systems?

If you ever need reminding that there may be a little more magic to our origins than current cosmology can offer, just remember the kiss of Venus, and the sage words of John Donne:

"Man hath weav'd out a net, and this net throwne
upon the Heavens, and now they are his owne.
Loth to goe up the Hill, or labour thus
to goe to Heaven, we make Heaven come to us."

Pleiades
Orion
Sirius
60°

SUN & PLANETS		Perihelion (10^6 km)	Mean Orbit (10^6 km)	Aphelion (10^6 km)	Eccentricity	Inclination of Orbit (degrees)	Perihelion Longitude (degrees)	Orbital Period (days)	Tropical Year (days)
Sun	☉	-	-	-	-	-	-	-	-
Mercury	☿	46.00	57.91	69.82	0.205631	7.0049	77.456	87.969	87.968
Venus	♀	107.48	108.21	108.94	0.006773	3.3947	131.53	224.701	224.695
Earth	+	147.09	149.60	152.10	0.016710	0	102.95	365.256	365.242
Mars	♂	206.62	227.92	249.23	0.093412	1.8506	336.04	686.980	686.973
Ceres	⚳	446.60	413.94	381.28	0.0789	10.58	???	1680.1	1679.5
Jupiter	♃	740.52	778.57	816.62	0.048393	1.3053	14.753	4,332.6	4,330.6
Saturn	♄	1,352.2	1,433.5	1,514.5	0.054151	2.4845	92.432	10,759.2	10,746.9
Chiron	⚷	1,266.2	2,050.1	2,833.9	0.38316	6.9352	339.58	18,518	18,512
Uranus	♅	2,741.3	2,872.46	3,003.6	0.047168	0.76986	170.96	30,685	30.589
Neptune	♆	4,444.4	4,495.1	4,545.7	0.0085859	1.7692	44.971	60,190	59,800
Pluto	♇	4,435.0	5,869.7	7,304.3	0.24881	17.142	224.07	90,465	90,588

MOONS (a selection)		Name of Satellite	Mean Orbital Radius (10^3 km)	Orbital Period (days)	Eccentricity of Orbit	Inclination of Orbit (°)	Diameter (mean) (km)	Mass (10^{18} kg)
Earth's	+	*The Moon*	384.8	27.3217	0.0549	5.145	3,475	73,490
Mars's	♂	*Phobos*	9,378	0.31891	0.0151	1.08	22.4	0.0106
		Deimos	23,459	1.26244	0.0005	1.79	12.2	0.0024
Jupiter's	♃	*Io*	421.6	1.7691	0.004	0.04	3,643	89,330
		Europa	670.9	3.5512	0.009	0.47	3,130	47,970
		Ganymede	1,070	7.1546	0.002	0.21	5,268	148,200
		Callisto	1,883	16.689	0.007	0.51	4,806	107,600
Saturn's	♄	*Tethys*	294.66	1.8878	<0.001	1.86	1,060	622
		Dione	377.40	2.7369	0.0022	0.02	1,120	1,100
		Rhea	527.04	4.5175	0.0010	0.35	1,528	2,310
		Titan	1,221.8	15.945	0.33	0.33	5,150	134,550
		Iapetus	3,561.3	79.330	0.0283	14.7	1,436	1,590

Rotation Period *(hours)*	Average Day Length *(hours)*	Equatorial Diameter *(km)*	Polar Diameter *(km)*	Axial Tilt *(degrees)*	Mass *(10^{24} kg)*	Volume *(10^{12} km^3)*	Surface Gravity *(m/s^2)*	Surface Pressure *(bars)*	Temp. (mean) *(°C)*
600 - 816	-	1,392,000	1,392,000	7.25	1,989,100	1,412,000	274.0	0.000868	5505
1407.6	4222.6	4,879.4	4,879.4	0.01	0.3302	0.06083	3.70	negl.	167
-5832.5	280.20	12,103.6	12,103.6	177.36	4.8685	0.92843	8.87	92	464
23.934	24.000	12,756.2	12,713.6	23.45	5.9736	1.08321	9.78	1.014	15
24.623	24.660	6794	6750	25.19	0.64185	0.16318	3.69	0.007	-65
9.0744	9.0864	960	932	var.	0.00087	0.000443	negl.	negl.	-90
9.9250	9.9259	142,984	133,708	3.13	1,898.6	1,431.28	23.12	100+	-110
10.656	10.656	120,536	108,728	26.73	568.46	827.13	8.96	100+	-140
5.8992	5.8992	208	148	unkn.	0.000006	0.000024	negl.	negl.	unkn.
-17.239	17.239	51,118	49,946	97.77	86.832	68.33	8.69	100+	-195
16.11	16.11	49,528	48,682	28.32	102.43	62.54	11.00	100+	-215
-153.29	153.28	2390	2390	122.53	0.0125	0.00715	0.58	negl.	-223

MOONS
(continued)

		Name of Satellite	Mean Orbital Radius *(10^3 km)*	Orbital Period *(days)*	Eccentricity of Orbit	Inclination of Orbit *(°)*	Diameter (mean) *(km)*	Mass *(10^{18} kg)*
Uranus's	♅	*Miranda*	129.39	1.4135	0.0027	4.22	235.7	66
		Ariel	191.02	2.5204	0.0034	0.31	578.9	1,340
		Umbriel	266.30	4.1442	0.0050	0.36	584.7	1,170
		Titania	435.91	8.7059	0.0022	0.14	788.9	3,520
		Oberon	583.52	13.463	0.0008	0.10	761.4	3,010
Neptune's	♆	*Proteus*	117.65	1.1223	0.0004	0.55	193	3
		Triton	354.76	-5.8769	0.000016	157.35	2,705	21,470
		Nereid	5,5413	360.14	0.7512	7.23	340	20
Pluto's	♇	*Charon*	19.6	6.3873	<0.001	<0.01	1,186	1,900

Only the major moons of the gas giants are given. There are currently 28 known moons around Jupiter, 30 around Saturn, 21 around Uranus and 8 around Neptune. There are probably many more.

DANCES OF THE PLANETS

MERCURY - VENUS

MERCURY - EARTH

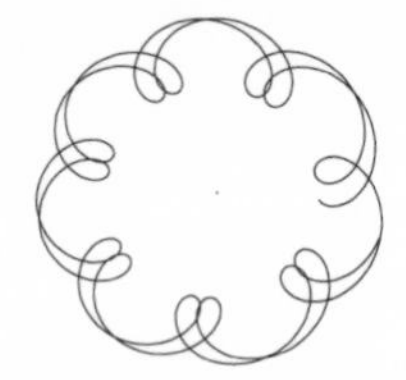

MERCURY - MARS

MERCURY - CERES

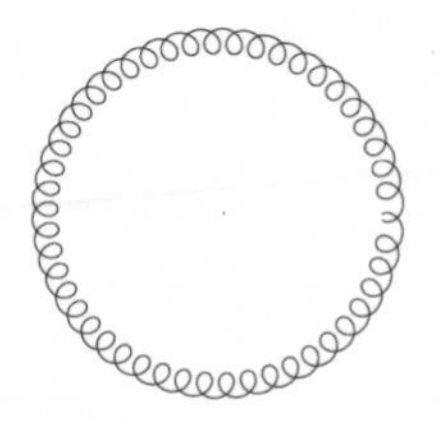

MERCURY - JUPITER

MERCURY - SATURN

VENUS - EARTH

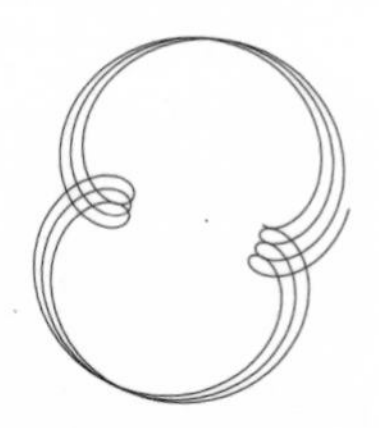

VENUS - MARS

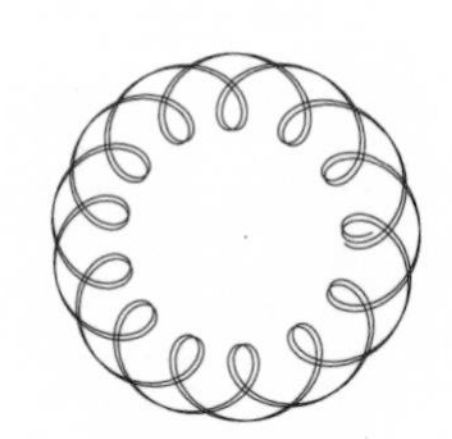

VENUS - CERES

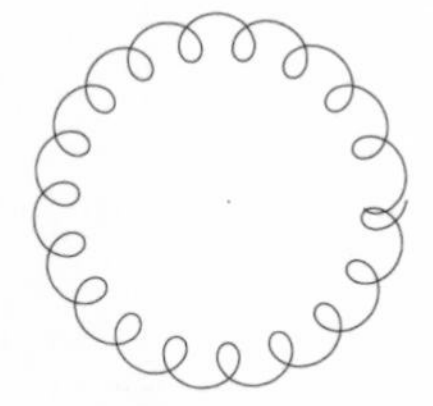

VENUS - JUPITER

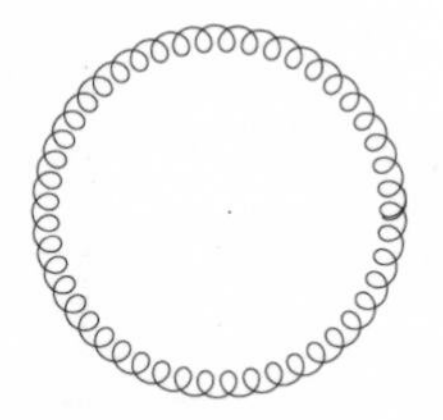

VENUS - SATURN

EARTH - MARS

EARTH - CERES

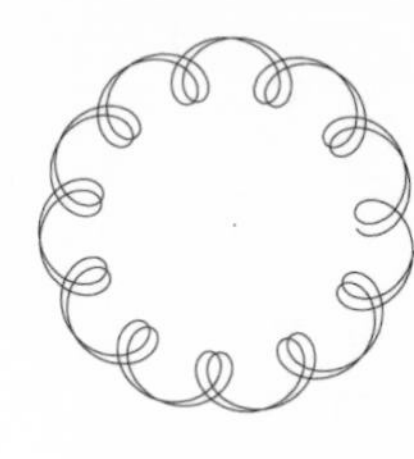

EARTH - JUPITER

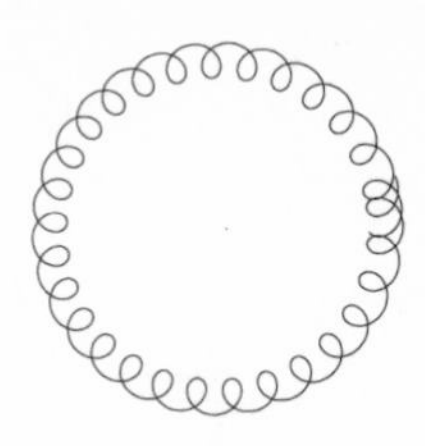

EARTH - SATURN

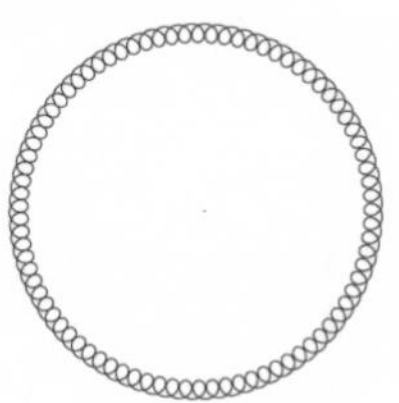

EARTH - URANUS

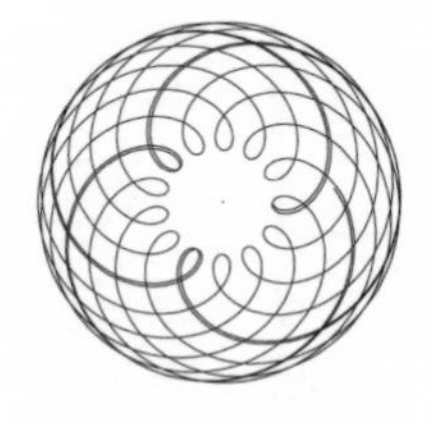

MARS - CERES

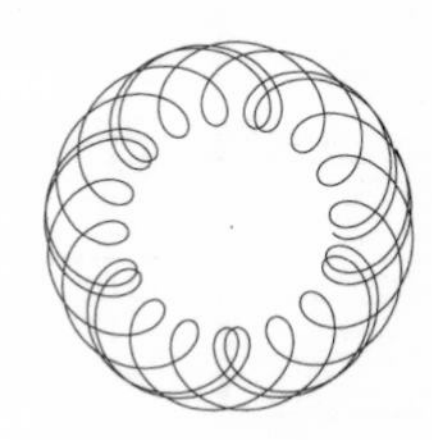

MARS - JUPITER

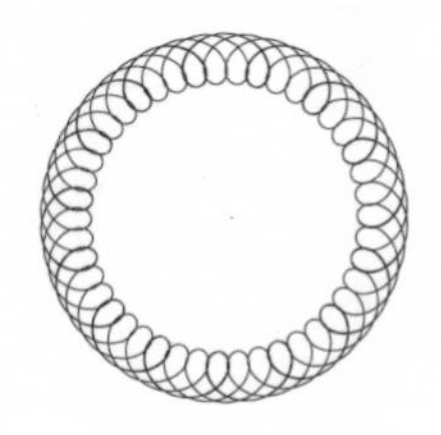

MARS - SATURN

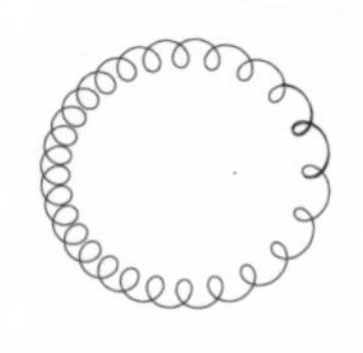

MARS - CHIRON

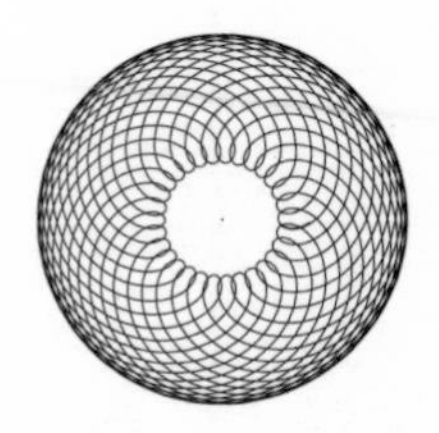

CERES - JUPITER

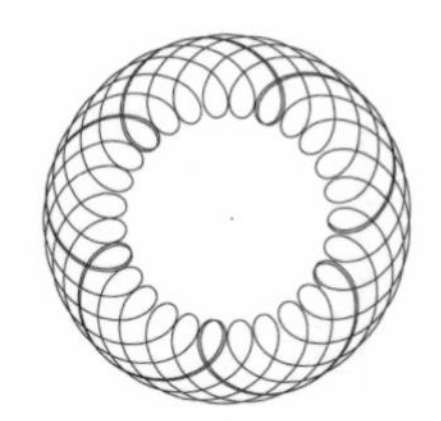

CERES - SATURN

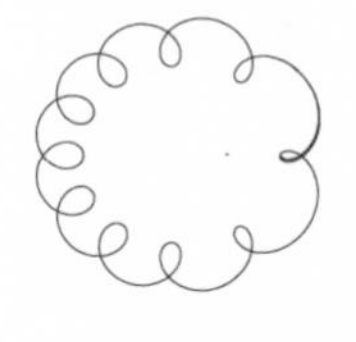

CERES - CHIRON

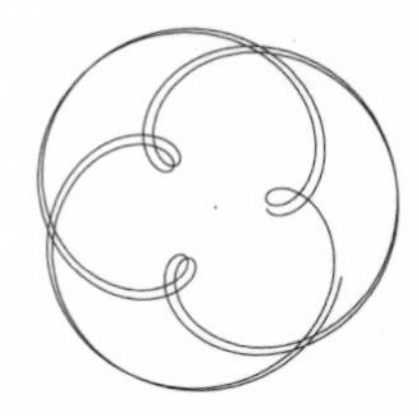

JUPITER - SATURN

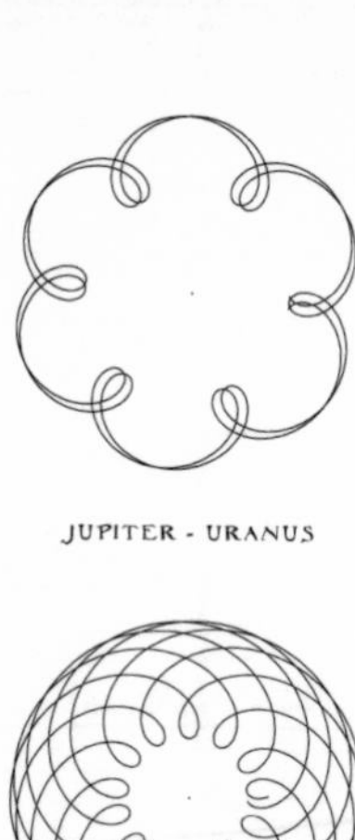

JUPITER - URANUS

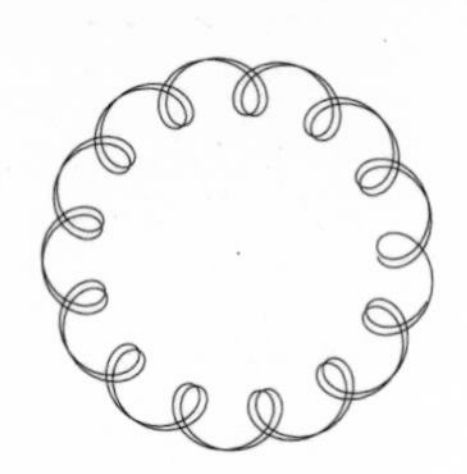

JUPITER - NEPTUNE

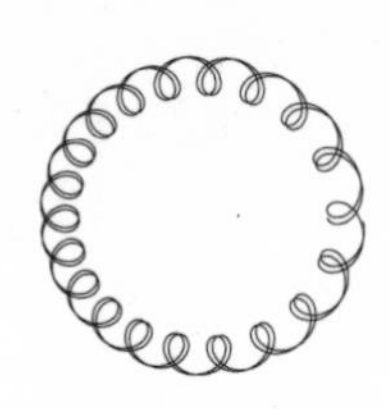

JUPITER - PLUTO

SATURN - URANUS

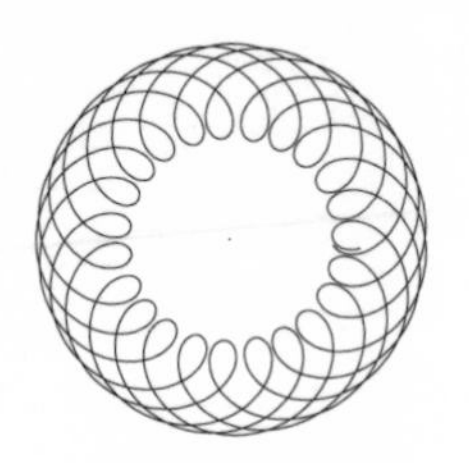

SATURN - NEPTUNE

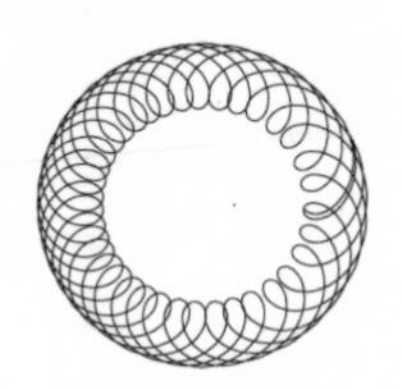

SATURN-PLUTO

CHIRON - URANUS

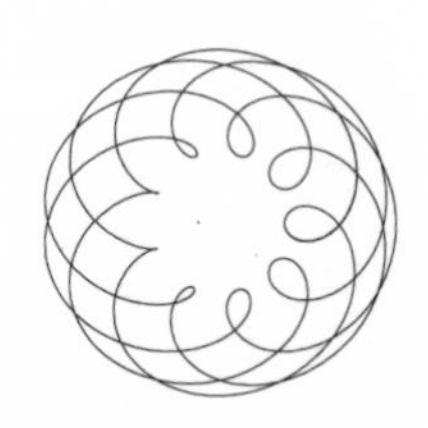

CHIRON - NEPTUNE

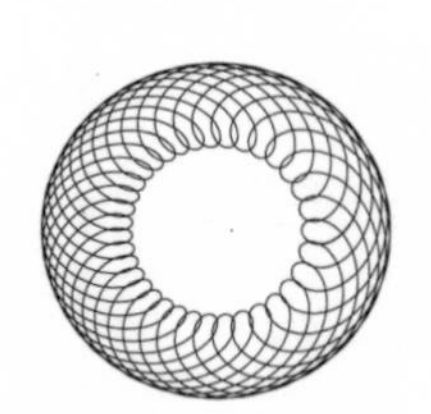

CHIRON - PLUTO

URANUS - NEPTUNE

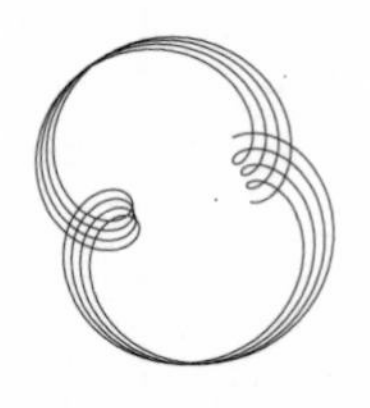

URANUS - PLUTO

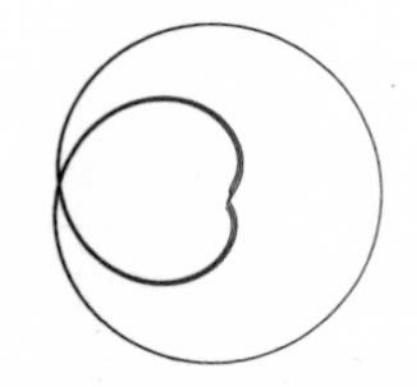

NEPTUNE - PLUTO